S0-BXB-684

PANBIOGEOGRAPHY

Oxford Biogeography Series
Edited by A. Hallam, B.R. Rosen, and T.C. Whitmore

Wallace's Line and Plate Tectonics
T.C. Whitmore

Cladistic Biogeography
Christopher J. Humphries and Lynne R. Parenti

Biogeography and Quaternary History in Tropical America
T.C. Whitmore and G.T. Prance

Biogeographical Evolution of the Malay Archipelago
T.C. Whitmore

Vanishing Rainforests: The Ecological Transition in Malaysia
S. Robert Aiken and Colin H. Leigh

Australian Rainforests
Paul Adam

The Africa–South America Connection
Wilma George and René Lavocat

The Palaeobiogeography of China
Yin Hongfu

Life History and Biogeography: Patterns in Conus
Alan J. Kohn and Frank E. Perron

An Outline of Phanerozoic Biogeography
Anthony Hallam

Panbiogeography: Tracking the History of Life
R.C. Craw, J.R. Grehan, and M.J. Heads

PANBIOGEOGRAPHY

Tracking the History of Life

Robin C. Craw

John R. Grehan

Michael J. Heads

New York Oxford

Oxford University Press

1999

Oxford University Press

Oxford New York
Athens Auckland Bangkok Bogotá Buenos Aires Calcutta
Cape Town Chennai Dar es Salaam Delhi Florence Hong Kong Istanbul
Karachi Kuala Lumpur Madrid Melbourne Mexico City Mumbai
Nairobi Paris São Paulo Singapore Taipei Tokyo Toronto Warsaw

and associated companies in
Berlin Ibadan

Published by Oxford University Press, Inc.
198 Madison Avenue, New York, New York 10016

Library of Congress Cataloging-in-Publication Data
Craw, R. C. (Robin C.)
Panbiogeography : tracking the history of life / Robin C. Craw, John R. Grehan,
and Michael J. Heads.
p. cm.—(Oxford biogeography series)
Includes bibliographical references and index.
ISBN 0-19-507441-6
1. Biogeography. 2. Life—Origin. I. Grehan, John R. II. Heads, Michael J.
III. Title. IV. Series.
QH84.C678 1999
578'.09—dc21 97-41638

9 8 7 6 5 4 3 2

Printed in the United States of America
on acid-free paper

Acknowledgments

Proofreading and general critique of manuscripts was kindly provided by James Boone, Scott Griggs, Trish Hanson, Jonathan Leonard, Judy Rosovsky, and Claudia Violette. Comments or information on specific topics were kindly contributed by Paul Ackery (British Museum of Natural History, London), Laurel Collins (Florida International University), Lucy Cranwell Smith (Tuscon, Arizona), Gabriel Dover (University of Leicester), John Elder (University of North Dakota), Tomas Feininger (Vieux-Quebec, Quebec), Keith Langdon (Great Smoky Mountains National Park), Paul Mann (University of Texas at Austin), Simon Mickleburgh (Fauna and Flora Preservation Society, UK), Brian Patrick (Otago Museum, New Zealand), Robert Ricklefs (University of Missouri-St. Louis), Rolf Sattler (McGill University), Jack Sites (Brigham Young University, Utah), Dwight Taylor, and Jonathan Wendel (Iowa State University). Particular thanks are also due to Ian Henderson (Massey University) for facilitating electronic transfers of manuscripts, Patricia Mardeusz and other library staff at the University of Vermont, Jerome Djangmah (University of Ghana), Adwoa Appiah, Wisdom Akonta, Martin Arkoh (University College, Winneba), everyone at Esi Adazewa Fie, Nozepo Nobanda, and Bob Drummond (National Herbarium, Zimbabwe) for their assistance.

Contents

PANBIOGEOGRAPHY

1

What Is Panbiogeography?

Recent disciplinary practice in biogeography has resulted in the formation of distinct approaches. The three approaches recognized by Vuilleumier (1978) were the theory of faunal regions and centers of origin, the methods of cladistics and vicariance, and island biogeography. Nelson (1978) loosely equated each of these approaches, respectively, with the major methods of biological systematics—evolutionary, cladistic, and phenetic. Although panbiogeography is often subsumed within the cladistic and vicariance approach (e.g., Nelson and Platnick 1981, Mayr 1982a, Stott 1984, Stoddart 1986, Briggs 1987, 1991), it was recognized as a distinct fourth approach by Hull (1988) in an important and insightful study of the development of evolutionary biology. This distinction has been confirmed by contemporary commentators (Bănărescu 1990, Myers 1990, Nelson and Ladiges 1990, Cranston and Naumann 1991, Wilson 1991, Cox and Moore 1993, Espinosa and Llorente 1993, Morrone and Crisci 1995, Morrone et al. 1996, Colacino 1997, Molnar 1997).

Anyone attempting to provide a sketch of panbiogeography has to contend with a curious diversity of portraits on display in the literature. A reading of this literature reveals enormous confusion over what exactly constitutes panbiogeography, its accomplishments, and its relationships to other areas of biogeography and different disciplines. To Grene (1990) panbiogeography is "faddish," while Mayr (1982a) finds it "eccentric." According to Stace (1989) it is "vicariance in worldwide distribution patterns." Wiley (1981) considers any insights of panbiogeography simply as repetitions of earlier work by Cain (1944) and Camp (1948), yet to Nelson (1989) it is a distinct evolutionary metatheory. So what is panbiogeography? Is it an evolutionary theory, an approach to biogeography, a conceptual synthesis, or a novel research program? Is it a crude precursor to vicariance biogeography (e.g., Grande 1990) or a rival of cladistic biogeography (e.g., Humphries and Seberg 1989, Platnick and Nelson 1989)? Is it some, none, or all of the above?

Panbiogeography is an attempt to reintroduce and reemphasize the importance of the spatial or geographical dimension of life's diversity for our understanding of evolutionary patterns and processes. It is an approach to biology that focuses on the role of locality and place in the history of life. Its goal is to recover the importance of places and localities as direct subjects of analysis in biogeography. Central to the panbiogeographic project is the acknowledgment that an understanding of locality is a fundamental precondition to any adequate analysis of the patterns and processes of evolutionary change. Panbiogeography emphasizes the role of place in the processes of the past as understood from the perspective of the present.

Unfortunately, many biogeographical, evolutionary, and systematic studies do not even map out geographical distributions of their target organisms. Phylogenies and ecologies are often portrayed as entities independent of geography. Geographical distribution becomes, at best, driftwood surfacing every now and then from seas of computer printouts and endless studies of phylogenetic trees. Most books about biogeography written by geographers emphasize the ecosystem concept rather than space, and contain remarkably little about distribution patterns. It is as if the geographic context was being written out of the subject altogether.

Biologists need to refocus on a sense of place and time as context. There are no ecological or evolutionary events that are not related in some essential way to the particular place and time of their occurrence. Spaces and places are not elements of evolution to be taken for granted. They represent a critical source of information on the natural units of organic distribution and therefore the biogeographical structure of the history of life. A richer conception of space is required in biology. The discipline of biogeography needs to see time in space and to see history unfolding in parts of the earth space—there is a necessary linkage between time and place. Geography is the substratum upon which the history of life takes place.

1.1 A Darwinian Dilemma

Modern evolutionary theory arose out of biogeography and ecology (Ghiselin 1980), and Darwin derived more evidence for evolution from the facts of biogeography than from any other biological phenomenon (Mayr 1982a). Darwin thought that biogeography was an interesting and, indeed, a critical subject, otherwise he would not have introduced his evolutionary theory to the world through the medium of this discipline. At this very point, where he knew that "certain facts in the distribution of organic beings" (Darwin 1859) provide evidence for the origin of species, Darwin recognized the critical importance of geography for evolutionary patterns and processes.

Darwin discussed biogeography in two chapters of *On the Origin of Species* (1859) in which he explained the phenomenon of vicariant distribution (i.e., the same or related forms of life inhabiting different localities) by a theory of historical descent. Many of Darwin's contemporaries had postulated independent multiple creations as an explanation for vicariant distributions. In Darwin's theory, however, they were the consequence of descent with modification. To counter the creationist view and to bridge the spatial and temporal gaps between vicariant distributions through descent from a common ancestor, Darwin appealed to a theory of chance migration from single and geographically restricted ancestral centers of origin. This center-of-origin concept embodies a general idea that ancestral species occupy restricted geographical ranges compared to those of their immediate descendants.

Darwin described a biological geography wherein organisms were constantly evolving and pouring out from centers of origin by means of dispersal (e.g., flying, walking, swimming, passive drifting, rafting). These different migratory abilities of organisms were seen by Darwin as representing a key mechanism of evolution in time and across space. The isolation, necessary for evolutionary differentiation, was provided by effective barriers against migration. Isolation occurred through ancestral founders crossing a preexisting geographic barrier or by migrating before the barrier came into existence. Mobility of organisms was seen as the key to their biogeography, and debates ensued over whether this movement was mediated by the active or accidental crossing of preexisting barriers (migrationist hypothesis) or by movement over former connections such as land bridges (extensionist hypothesis).

Sir Joseph Dalton Hooker, an acknowledged authority on plant geography, a leading botanist, and a trusted colleague of Darwin's, was caught up in this dilemma of conflicting migration and land extension hypotheses as biogeographical explanations. Early in his career (Hooker 1853), he showed a marked preference for extensionist hypotheses to explain major disjunctions in distribution in the Southern Hemisphere (e.g., the southern beeches *Nothofagus* distributed in Australasia and southern South America), but Hooker's general attitude to migrations and barriers became equivocal. Following publication of Darwin's *Origin of Species*, Hooker began to give greater emphasis to the role of migration, but he persisted in his doubt that either explanation was sufficient for understanding evolution in space and time.

Very early in the debate, Hooker realized that either hypothesis could fully explain any given distribution, so choosing one over the other did not provide a real solution (Darwin and Seaward 1903). This conflict over the relative importance of migration or land extension may be regarded as an early version of the dispersal (= migration) versus vicariance (= splitting of distributions following connections through land bridges or drifting continents) dispute recently highlighted in works devoted to vicariance biogeography (e.g., Nelson and Rosen 1981, Wiley 1988). Hooker suggested that the solution of this dilemma required the formulation of a different kind of reasoning and method that did not exist in his time. Panbiogeography represents a potential solution to this puzzle because it does not postulate any preconceived position on dispersal or vicariance as biogeographical explanations. Instead it is concerned with establishing a general relationship between the dispersal of organisms and the vicariance of taxa.

1.2 Panbiogeography, Dispersal, and Vicariance

Means of Dispersal and Geographic Distribution

Much of the history of biogeography has been dominated by proposals that means of dispersal are the principal factor responsible for the evolution of distribution. Organisms can and do move about according to their means of dispersal, and through chance events they can colonize new territories and habitats, whether by crossing geographic barriers or from dispersion across paleogeographical features such as land bridges or joined land masses before rifting and drifting. Different means of dispersal in differently dispersing taxa should result in mutually incongruent distribution patterns, but there is no consistent relationship between means of dispersal and the extent of the geographical distribution of a group. Taxa known to have good means of dispersal may be widely or narrowly distributed, and taxa with poor or no obvious means of dispersal may have extensive geographical ranges.

Mayr (1982b) has claimed that poor dispersers such as earthworms and primary freshwater fish have totally different distribution patterns from winged birds and butterflies, but this contrast does not always prove to be true. Danaine butterflies (*Tirumala*) and primary freshwater Sisorioidea fish, for example, share a similar distribution range (fig. 1-1). These two groups show some local differences—the fishes occurring in the Middle East, and the butterflies in the western Pacific—but their distributions are not distinguishable according to their different means of dispersal.

The proposition that different means of dispersal take precedence over actual distribution was argued forcefully by Haydon et al. (1994a: 404):

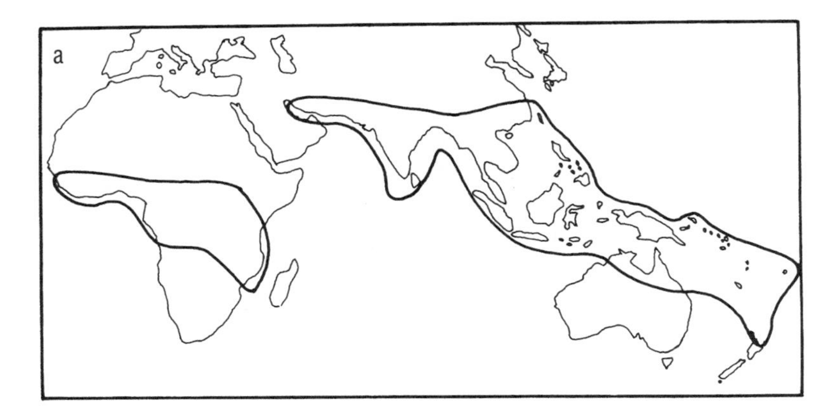

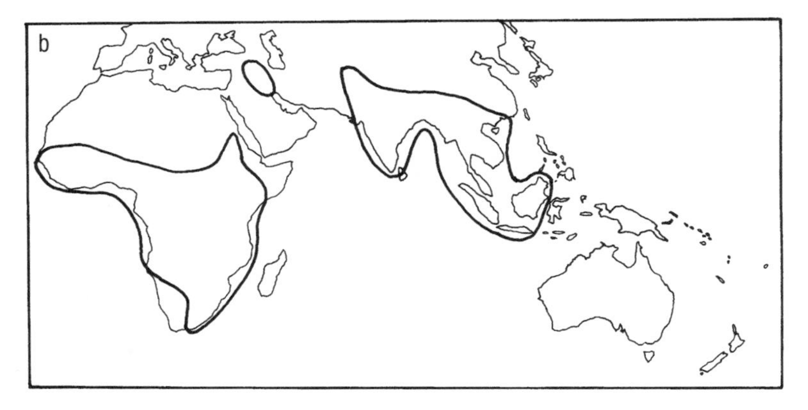

FIGURE 1-1. Comparison of generalized distribution patterns for (a) the danaine butterfly genus *Tirumala* (Ackery and Vane-Wright 1984) and (b) a group of primary freshwater fish (Superfamily Sisorioidea) (Bănărescu 1990). Despite their different means of dispersal, the distributions of these two groups are geographically similar. This similarity of distribution casts doubt upon Mayr's (1982b) supposition that such organisms have "totally different" distributions.

> Bats and spiders possess powers of dispersal that allow them to alter their distributions much more readily than can land snails or fossorial blind snakes. Therefore, the biogeography of bats and spiders confounds attempts to recover general patterns and responsible processes. Snails and blind snakes provide a better historical signal because they are less capable of altering the geographic distribution dealt them by vicariant processes.

There is equivocal evidence supporting this contention that snails and blind snakes are necessarily "better" historical biogeographical

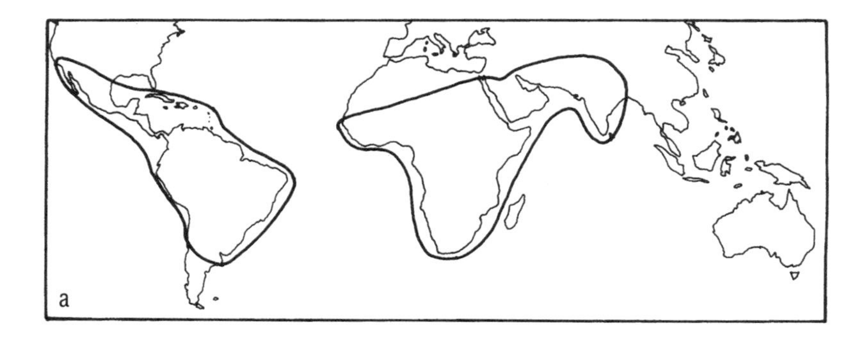

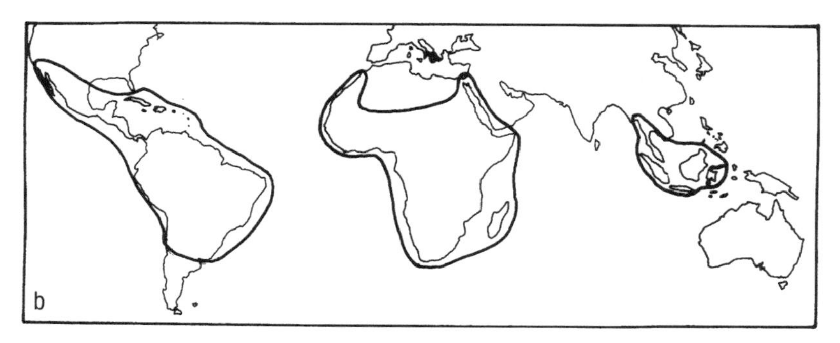

FIGURE 1-2. Comparison of the generalized distribution of (a) blind snakes, Leptotyphlopidae, (after Terent'ev 1965) and (b) leaf-nosed vampire bats (Phyllostomatidae) and hispid bats (Nycteridae) (after Koopman 1984). Note the near identical distribution range of snakes and bats in the Americas, and the extensive range of snakes and bats in the Old World.

subjects than bats or spiders. For instance, bats and blind snakes can show congruent biogeographical patterns. The distribution range of fossorial blind snakes in the Americas is nearly identical to that of the bat family Phyllostomatidae, while in the Old World these snakes have a distribution that is almost as extensive as that of the bat family Nycteridae (fig. 1-2).

Biogeographical congruence between bats and snails is illustrated by the distributions of a land snail genus *Bocageia* and subspecies of the bat *Miniopterus minor* in Africa and neighboring islands (Gascoigne 1994). The snail genus occurs on the island of São Tomé (Gulf of Guinea) and the Comoro Islands (off East Africa) and in Central Africa (on the Ruwenzori range), while the bat likewise has several subspecies distributed between the islands of São Tomé and Madagascar (fig. 1-3). They share this Gulf of Guinea–East African/western Indian Ocean distribution pattern with many other groups of organisms (Croizat 1952, 1958; Aubréville 1974b, Bramwell 1990).

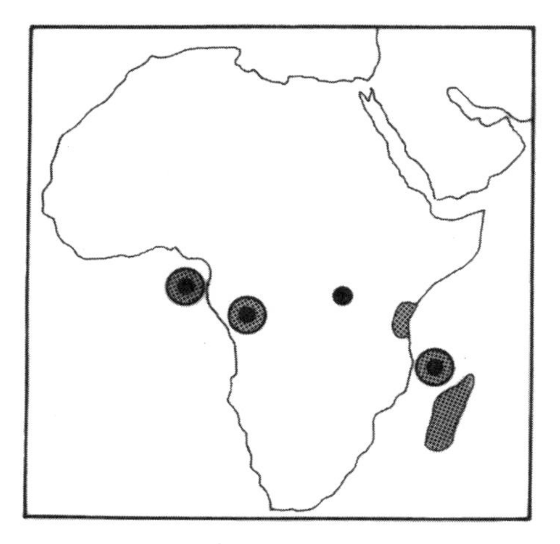

FIGURE 1-3. Distribution of the land snail *Bocageia* (filled circles) on São Tomé (Atlantic), and the Comoros (Indian Ocean) (after Gascoigne 1994) compared with subspecies (one for each stippled area) of the bat *Miniopterus minor* (after Juste and Ibáñez 1992). Note that these organisms with very different means of dispersal share a similar geographic pattern of distributional disjunctions.

It is true that some bat species will be more widespread and vagile than others, and this is also true of snails, spiders, butterflies, freshwater fishes, and blind snakes, but these variations in distributional range do not exemplify absolute differences in biogeographical pattern or historical signal. Differing dispersal abilities of different organisms do not delineate fundamentally different distribution patterns. Consideration of them cannot, therefore, qualify as a necessary prerequisite to the analysis of distribution patterns. In fact, the converse may be true—pattern analysis may provide critical insight into the significance and impact of dispersal ability on the origin of distribution.

Means of Dispersal versus Vicariance

Hypotheses explaining distribution by casual migration through means of dispersal out of centers of origin lead to models of evolution and differentiation involving founding individuals or populations that fortuitously cross barriers and successfully colonize new territory or habitat. These founders represented rare chance events involving a small number of individuals so that the lineage is forced through a genetic bottleneck. Founder events have been theorized to represent an important mode of speciation as the genetically depauperate population adapts to the new habitat, but an experimental test on *Drosophila* populations sug-

TABLE 1-1. Examples of studies of selected taxa where vicariance has been postulated as a major determinant of patterns of geographical distribution.

Taxonomic group	Source
Algae, Rhodophyta (Gigartinaceae, Petrocelidaceae, Phyllophoraceae)	Guiry and Gabary 1990
Lichens, Roccellaceae	Galloway 1988, Tehler 1983
Filicales, Polypodiaceae (*Pyrrosia*)	Hovenkamp 1986
Coniferae, Pinaceae	Malusa 1992
Angiosperms, seagrasses	Fortes 1988
Dipterocarpaceae	McGuire and Ashton 1977
Myrtaceae (*Eucalyptus*)	Ladiges and Humphries 1986
Portulaceae	Carolin 1987
Rosaceae (*Crataegus*)	Phipps 1983
Sapindaceae	Van Welzen et al. 1992
Solanaceae	Knapp 1991
Porifera, Demospongiae, Chalinidae	De Weerdt 1989
Coelenterata, Hydrozoa, (*Millepora*)	De Weerdt 1990
Nematoda, Leptanchoidea	Ferris 1979
Turbellaria, Tricladida, Paludicola	De Vries 1985
Platyhelminthes, Cercomeria, Cercomeromorpha	Bandoni and Brooks 1987
Eucestoda, Tetrabothriidae	Hoberg and Adams 1992
Oligochaeta, Megascolecidae	Jamieson 1981
Crustacea, Amphipoda	Lindemann 1990, Stock 1993
Copepoda, Harpocticoida	Ho 1988
and hydroids	Cunningham et al. 1991
Arachnida, Araneae	Platnick 1976
Scorpiones, Iuridae	Francke and Soleglad 1981
Insecta, Coleoptera, Amphizoidae	Peiyu and Stork 1991
Carabidae, Oodini	Spence 1982
Carabidae, Caelostomina	Liebherr 1986
Carabidae, Platynini	Liebherr 1986
Carabidae, Pterostichini	Allen 1980
Dytiscidae	Baumel 1989
Dytiscidae	Wolfe 1985
Dytiscidae	Wolfe and Roughley 1990
Hydraenidae	Perkins 1980
Chrysomelidae	Askevold 1991
Curculionidae	Spanton 1994
Insecta, Diptera, Chironomidae	Cranston and Edwards 1992
Keroplatidae	Matile 1990
Insecta, Ephemeroptera	Bae and McCafferty 1991
Insecta, Hemiptera, Veliidae	Anderson 1989
Insecta, Heteroptera, Nabidae	Asquith and Lattin 1990
Miridae: Orthotylinae	Asquith 1993
Insecta, Hymenoptera, Vespidae	Cumming 1989
Insecta, Isoptera	Salick and Pong 1984
Insecta, Lepidoptera, Gelechiidae	Pitkin 1988
Nymphalidae	Miller and Miller 1990
(*Boloria*)	Britten and Brussard 1992
Papilionidae	Miller 1987, Hammond 1991

TABLE 1-1. *(continued)*

Taxonomic group	Source
Insecta, Odonata	Battin 1992
Insecta, Trichoptera, Triplectinidae	Holzenthal 1986
Echinoidea, Clypeasteroidea	Harold and Telford 1990
Echinodermata, Asteroidea	Rowe 1985
Mollusca, Gastropoda, Olividae	Michaux 1991
Pulmonata	Rousseau and Puisségur 1990
Actinopterygii, Lepisosteidae	Wiley 1976
Pisces, Characiformes	Vari 1989
Pisces, Curimatidae	Vari 1991
Pisces, Teliostei, Cyprinidae	Howes 1984
Engraulidae	Nelson 1984
Gadoidei	Howes 1990, 1991
Poeciliform fishes	Rosen 1978
Blennioidei	Stepien 1992
Cyprinodontidae	Wiley 1977
Pisces, Perciformes, Pseudochromidae	Winterbottom 1986
Pisces, Salmoniform fishes	Rosen 1974
Pisces, Percoidei, Teraponidae	Vari 1978
Anura	Duellman 1986
Anura, Hylidae	Cannatella 1980, Duellman and Campbell 1992
Anura, Pipidae, (Pipa)	Trueb and Cannatella 1986
Anura, Leptodacylidae	Duellman and Veloso 1977, Heyer and Maxson 1982
Archosauria, Ornithischia	Milner and Norman 1984
Reptilia, Crocodilia and their digenean parasites	Brooks 1979
Reptilia, Gekkonidae	Leviton and Anderson 1984, Bauer 1990
Reptilia, Sauria, Leiolopisma	Zug 1985
Aves, Gruiformes	Cracraft 1982
Aves, Phalacrocoracidae	Siegel-Causey 1991
Mammalia, Rodentia, Muridae	Musser and Holden 1991, Rickart and Heaney 1991
Sciuridae	Patterson 1982, Lindsay 1987
Insectivora	MacFadden 1980

gests that new species are an unlikely by-product of such events (Moya et al. 1995).

Theories based on casual dispersal and centers of origin remain common modes of biogeographical explanation. However, many recent studies have postulated vicariance as the primary biogeographical process in the evolution of individual animal and plant groups (table 1-1) and communities (table 1-2). A comparative study of the prevalence of vicariance in the evolution of different groups led to the conclusion that for 66 examples of fishes, frogs, and birds, speciation was primarily due

TABLE 1-2. Examples of community and regional biogeographical studies where vicariance is postulated as a major determinant of geographical distribution patterns.

Community or region	Source
Australian Anura	Watson and Littlejohn 1985
Australian avifauna	Cracraft 1986
Baja California reptiles	Murphy 1983a,b
Bipolar molluscan disjuncts	Crame 1993
Caribbean fauna	Rosen 1976, Poinar and Cannatella 1987
Caribbean freshwater fauna	Harrison and Rankin 1976, Flint 1978, Saether 1981
Central Pacific marine mollusca	Rehder 1980
Chinese freshwater fish	Zhao 1991
Hydrothermal-vent fauna	Tunnicliffe 1988
Indo-Pacific reef corals	Pandolfi 1992
Marine faunal provinciality	Hayami 1989
Mediterranean insects	La Greca 1990
Mediterranean vertebrates	Oosterboek and Arntzen 1992
Montane mammals of New Mexico	Patterson 1980
Pacific reef fishes	Heads 1983
Palaeobiological communities	Boucot 1990
Polar biotas	Crame 1992
Southern Hemisphere biotas	Cracraft 1975
Southwestern Pacific biotas	Cracraft 1980
Tropical strand plants	Kolbek and Alves 1993

to vicariance (through allopatric and peripheral isolation models) in 86% of cases, while sympatric models accounted for 6% of cases. Dispersal was considered to be an unnecessary assumption in speciation studies (Lynch 1989). Chesser and Zink (1994) thought that the incidence of sympatric speciation was underestimated in this study, but they concluded that vicariance was the most prevalent mode of speciation.

Recent biogeographical discussion has been dominated by these and similar attempts to decide between dispersal and vicariance (e.g., Brignoli 1983, Poynton 1983, Thornton 1983, Ghiold and Hoffman 1984, Leis 1984, Schoener and Schoener 1984, Rosenblatt and Waples 1986, Lynch 1989, Stace 1989, Turner 1991). Some have suggested that the choice is ultimately untestable (Ghiold and Hoffman 1984, Leis 1984), and this position reflects Hooker's original view that one biogeographic possibility does not eliminate the alternative. Given that an organism is distributed over parts of the former supercontinent Gondwana, is it dispersal or vicariance that explains its presence in Africa and South America if the ancestors of that organism "crossed" the Atlantic Ocean before it was even a stream at the bottom of a rift valley? And what does it matter whether these ancestors crossed the "Atlantic" when it was only 1, 10, or 100 km wide? Is there any real biogeographical difference?

The vicariance–dispersal dichotomy requires organisms to belong to an "area" on one or the other side of a "barrier." This focus may be contrasted with the question of why an organism or taxon occurs within a particular geographic setting and not any other. A bird or moth arriving in a new habitat after crossing an intervening sea may be presumed to represent a "dispersal" event if the sea is treated as a barrier isolating separate biogeographic areas, domains, or regions. If instead the sea is considered a biogeographic focus for the bird or moth (i.e., the geographical history of the sea basin may have some relationship with the bird's or moth's present distribution), the geographical context of this dispersal event is the same as that of an organism "crossing" the same sea standing or sitting on some piece of "drifting" rock or continent. In this spatial setting the arrival is no more "vicariance" than "dispersal"—the focus of geographic context and evolutionary history remains the same. This comparison places ecological dispersal against a biogeographical background rather than endlessly opposing vicariance and dispersal.

Panbiogeographic Resolution

Different organisms with varying means of dispersal share similarities in their distribution patterns. There is nothing in plant geography that is not repeated by animals in some form or another. These similarities in distribution patterns suggest that different taxa may share a common biogeographic history by virtue of their ancestors occupying the same paleogeographic sector and being subjected to the same geological, geomorphological, and climatic changes. In this context means of dispersal are not the key to the present occurrence of taxa in different localities and places, but they are central to the formation of ancestral distribution ranges (Croizat 1993). If related taxa occupy different areas without having individually dispersed into them, it is necessary for their ancestors to have occupied those areas in the first place.

Vicariant form-making is the evolutionary differentiation of taxa *in situ*. The term "vicariant" refers to the spatial replacement of one taxon by another closely related one, even though they may often be geographically adjacent or even overlap to a limited extent. "Form-making" refers to the structural, developmental, physiological, and genetic evolution of organisms as distinct from their spatial and temporal histories.

The similarities and differences between dispersal, vicariance, and vicariant form-making models can be summarized in a diagram (fig. 1-4). In the dispersal model the emphasis is on the fortuitous crossing of preexisting barriers by ancestral organisms according to their different means of dispersal (fig. 1-4a). The vicariance model emphasizes the formation of a barrier after ancestral dispersal has taken place. This barrier

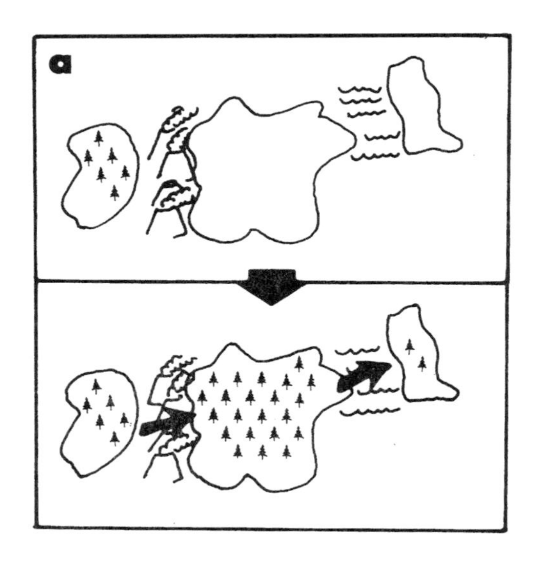

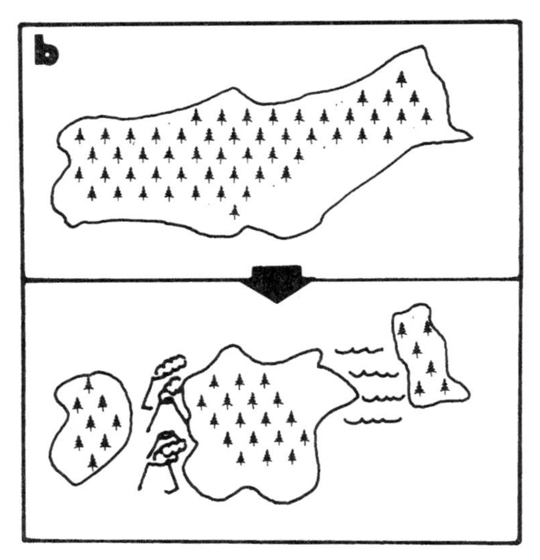

FIGURE 1-4. Biogeographic models of dispersal and form making. (a) Organisms or taxa reach new territory by crossing preexisting barriers (dispersalist biogeography); (b) organisms or taxa reach new territory by floating continents or land bridges before a barrier forms (vicariance biogeography) (after Savage 1982); (c) organisms or taxa reach new territory during a mobilist phase with geographic variation already present (*a*, *b*), followed by immobilist phase of differentiation of taxa correlated with subsequent geological and geomorphological events (*a′*, *b′*, *a1-3*, *b1-3*) (after Croizat 1964).

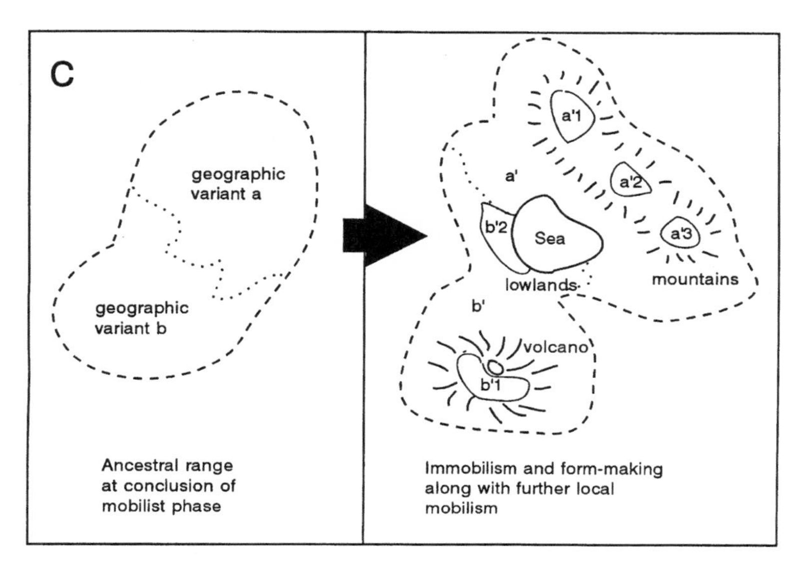

FIGURE 1-4. *(continued)*

then provides the isolation necessary for the subsequent differentiation of taxa (fig. 1-4b). Vicariance models postulate the splitting of ancestral ranges through the formation of barriers, whereas dispersal models presume the formation of ancestral ranges by the crossing of barriers.

Both the dispersal of ancestral organisms and the vicariant differentiation of ancestral populations are necessary for biological and geographical evolution (Craw 1978a). The vicariant form-making model emphasizes a widespread ancestral range, and the possibility that this ancestral population is already geographically differentiated is acknowledged (fig. 1-4c). Taxic differentiation occurs in relation to geographic and topographic alterations that have influenced or promoted but not predetermined the vicariant form-making of taxa. Isolation becomes a relative term referring to the differing spatial relations between ancestors and descendants rather than a phenomenon dependent on an external barrier.

Vicariant Form-Making, Dispersal, and Ecology

The vicariant form-making model distinguishes between two important phases in the evolution of a distribution. Establishment of an ancestral range requires that the ancestral organisms disseminate over a new geography. In this situation organisms are geographically mobile, and this biogeographic phase is termed "mobilism." This mobility of ancestors is mediated by means of dispersal—the different locomotory mechanisms of organisms (e.g., walking, flying, drifting). Thus, the ability of an ancestral organism to establish on new territory through its means of dispersal results in expansion of geographic range. Mobilist phases alternate with periods in which the ancestral range is stable, and no significant range expansion occurs. The locomotory abilities of the organisms persist, but they function as the means by which different sexes come into contact, or offspring occupy viable habitat within the established range (e.g., young animals moving from parental nests, plant seeds deposited on unoccupied substrate). If the landscape beneath ancestral populations is changing (e.g., through uplift, erosion, subduction), these means of dispersal will be critical in maintaining the survival of the evolving taxa as local habitats are either enhanced, degraded, expanded, or eliminated. This ecological and biogeographic relationship highlights the fallacy of treating means of dispersal as mysterious (e.g., van Steenis 1936, Haydon et al. 1994b).

Means of dispersal have two different biogeographic contexts in the vicariant form-making model instead of representing a biological constant. This contextual approach to means of dispersal renders "dispersal" a problematic term. In biogeography the meaning of the term "dispersal" is borrowed from ecology and refers solely to the migration and movement of organisms and taxa, so that the role of "dispersal" and "means of dispersal" is synonymous. In biogeography the problem of dispersal concerns the origin of distributions. Vicariant differentiation results in different spatial patterns as if actual migration of the vicariant descendants had occurred. Through vicariant form-making, the boundary and character of biogeographic space has been changed or translated from one set of forms (ancestors) to another (descendants). This translation in space through vicariant form-making represents the biogeographic concept of dispersal, which includes both mobilist and immobilist phases (Croizat 1964).

Alternation of mobilism and immobilism may be characterized in terms of ecesis—the ability of an organism to successfully establish in a new environment. During immobilism ecesis may be restricted as seeds, for example, may have little room for establishment beyond a narrow geographical range determined by a particular ecological association between organism and environment. Dispersal of these seeds contributes only to the means of survival for the organisms, and only a few may germinate and survive to produce more seeds. Under mobilist conditions many seeds may contribute to the establishment of new plants, and a fast rate of geographic expansion may occur. The seeds now constitute a means of dispersal, and a former relic or narrowly distributed endemic may now become an aggressive and prolific weed ("weed" is used here to refer to plants and animals that are highly vagile and able to establish prolifically over a particular area). This condition of weediness is not absolute, but it is a biogeographic function of place, time, and organism. The contingent role of dispersal suggests that the ability to disperse lies neither with organisms, nor is it consequent upon environmental changes, but lies in their reciprocal relationship specific to particular times and places.

The vicariant form-making model can be summarized as involving alternating cycles of dispersal and vicariance:

1. Ancestral widespread forms (mobilism),
2. Vicariance of widespread ancestral forms resulting in less widely distributed descendent forms (immobilism),
3. Dispersal of descendent forms producing further widespread forms (mobilism),
4. Vicariance of descendent widespread forms resulting in further less widely distributed descendent forms (immobilism).

Case Study: Admiral Butterflies

Mobilism and immobilism represent a biogeographic context within which means of dispersal may be evaluated. Differing degrees of sympatry and allopatry in spatial patterns may suggest different phases of mobilism. The distribution of admiral butterflies, for example, includes several overlapping ranges, although most of the species are vicariant to one another (fig. 1-5a). Geographic and phylogenetic relationships between these species suggest that their distributions can be interpreted as an underlying vicariant pattern formed in immobilism upon which is superimposed several subsequent dispersal or mobilist events (fig. 1-5b). Before across-barrier dispersal can take place, a species must have become a species through speciation, and if speciation is allopatric, then vicariance is prior or primary to the occurrence of dispersal.

By focusing solely on dispersal events in any particular case, the vicariant patterns may be overlooked, but by studying the vicariant patterns, the known cases of across-barrier dispersal in a particular group can be placed in a broader biogeographic perspective. The barrier concept is problematic because a barrier can only be defined in relation to an organism, so there are no independent criteria for specifying, and hence testing, the role of a barrier. Any distributional limit may be attributed to a barrier because a barrier is, by definition, a feature correlated with the limits of a distribution. This concept of a barrier may be less critical to biogeography than establishing the geographic context of mobilist and immobilist phases in the evolution of distributions. Both means of dispersal and vicariance are recognized in panbiogeography as significant biogeographical processes, but neither takes precedence over the analysis of distribution patterns.

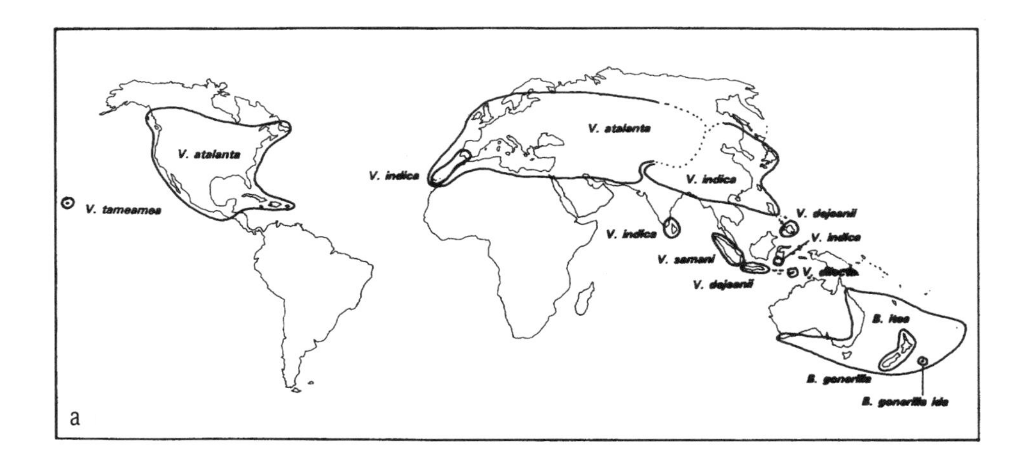

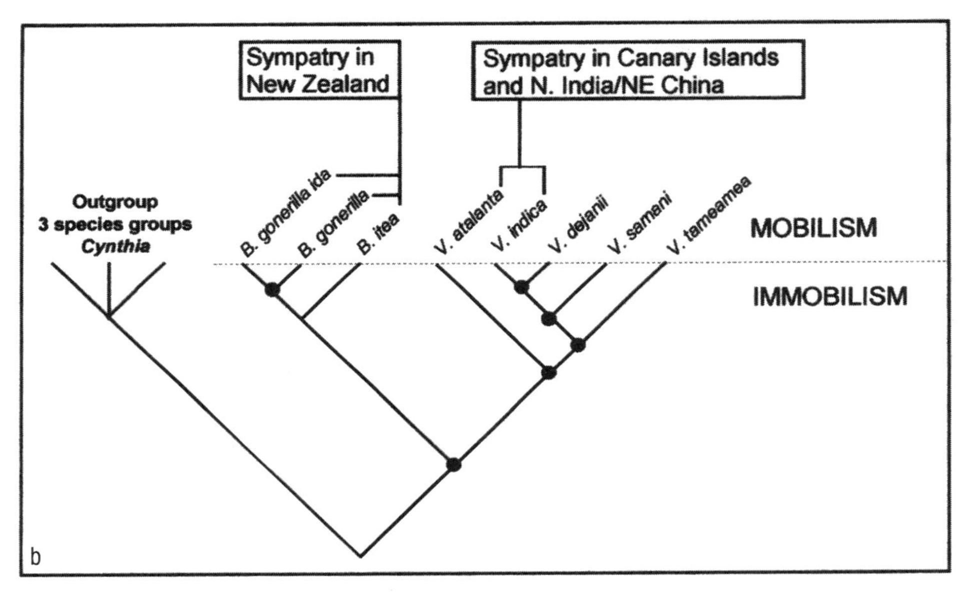

FIGURE 1-5. Distribution, dispersal, and vicariance in the nymphalid admiral butterfly genera *Bassaris* and *Vanessa*. (a) Generalized distribution ranges of vicariant and sympatric species (data from Field 1971, Leestmans 1978, Hanafusa 1992, P. Ackery, personal communication); (b) distributions interpreted as an underlying pattern of geographical vicariance (solid black circles) upon which is superimposed "noise" caused by secondary, across-barrier, dispersal events (from Craw 1990; reprinted with permission from SIR Publishing). (In panel a, species *V. dilecta* described by Hanafusa (1992) was not included in Craw's [1990] cladogram.)

1.3 Panbiogeographic Method

Panbiogeography not only attempts to resolve the dispersal–vicariance opposition, but it also offers a method of analysis for comparing geographic distribution patterns. The method requires a different set of assumptions to those of dispersalist and vicariance biogeography, including:

1. Distribution patterns constitute an empirical database for biogeographical analysis;
2. Distribution patterns provide information about where, when, and how animals and plants evolve;
3. The spatial and temporal component of these distribution patterns can be graphically represented, and;
4. Testable hypotheses about historical relationships between the evolution of distributions and earth history can be derived from geographic correlations between distribution graphs and geological/geomorphic features.

Every new method and theory brings with it the burden of novel terminology, or new applications of well-established terms. Panbiogeography is no exception. New terms can become sources of controversy as practitioners attempt to reconcile contrasting methods and interpretations. Many problems with terminology are not simply a matter of adequate definition, but a consequence of general familiarity with the context in which those terms are applied. For example, many standard cladistic terms and definitions that are now commonplace were widely contested or debated when cladistics was being defined and refined in the 1970s.

A panbiogeographic focus on patterns of distribution requires conceptual and methodological tools that allow comparisons to be made in a meaningful and informative way. Any two distributions can be compared, but specific biogeographical criteria are necessary to identify distributions that exhibit the same spatial characteristics (homologous distributions) and those that do not (nonhomologous distributions). In panbiogeography comparisons are spatial, and the concept of homology is linked to geographic characters that allow for the correlation of distribution patterns with patterns of geology, geomorphology, or tectonics, to formulate hypotheses of a common historical relationship. Analysis of distribution patterns and spatial homology relationships is called the track method and involves four principal concepts: track, node, main massing, and baseline (fig. 1-6).

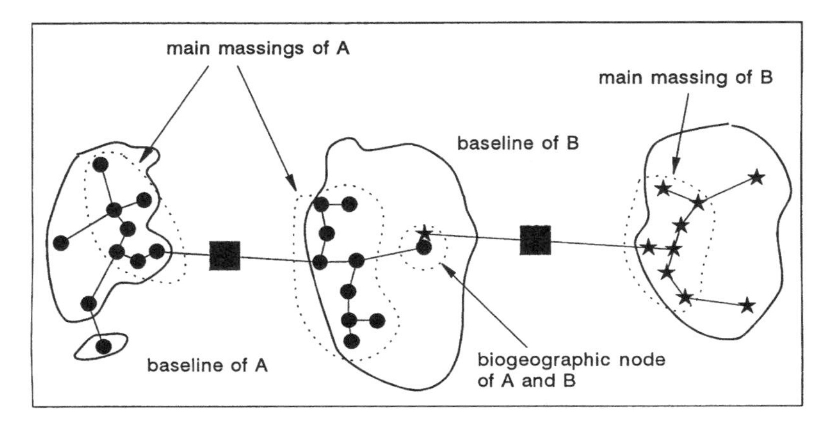

FIGURE 1-6. Generalized conceptual diagram of terms used in the panbiogeographic method. Enclosed areas represent geographic areas separated by ocean or sea basins.

Track

A track is a line drawn on a map that connects the different localities or distribution areas of a particular taxon or group of taxa. The simplest way to construct such a graph is to form a minimal spanning tree. This is an acyclic graph that connects all localities or distributional areas for a particular taxon or subordinate taxa belonging to a higher taxon (e.g., species of a genus, genera of a family), such that the sum of the link lengths connecting all the localities or distributional areas is the smallest possible.

These lines explicitly give shape or expression to the space and time that necessarily intervenes between disjunct localities. The immediate implication of this representation is that the evolution of distributions involves not only the form comprising the individual organisms at each locality or area, but also space and time. As a general principle, the evolution of a distribution is a function of form, space, and time together. The drawing of a track identifies the spatiotemporal coordinates most closely associated with the origin of that distribution. The shape or geometry imparted by these coordinates can then be compared to those of other patterns, and hypotheses or predictions about their origin proposed (e.g., specific historical events or processes).

Node

A node is the point of intersection between two or more tracks. Nodes are the point corollary of the line track. Where tracks connect different localities, nodes serve to delineate biogeographical boundaries. In this role, the concept of boundary is no longer a line encircling a distinct area of endemism, but a relationship between two or more tracks. Just as there are standard tracks composed of individual tracks, there are standard nodes composed of multiple individual nodes involving different tracks. The relative importance of individual nodes can be the subject of biogeographical analysis about the number and types (homology) of tracks involved, and how they interact either by intersecting, or reaching their peripheral limits at the node (see chapter 5). Nodes are biogeographically interesting because they are composite structures that imply a suturing of different biogeographical and ecological histories.

A node, as a boundary or intersection between different tracks, may exhibit one or several significant biological characteristics such as local presence (endemics), local absence (of taxa widespread or dominant elsewhere), diverse phylogenetic and geographic affinities, and the geographic or phylogenetic limits of taxa (Heads 1990). Important nodes are more likely to exhibit most, if not all, these characteristics. The more prominent nodes are sometimes called "gates" because they are points connecting past history with present-day distributions. Most biogeographers usually focus their attention on the presence of taxa, but the absence of taxa can also be informative about the significance and origins of distribution, and this absence is designated by the term "antinode" (Henderson 1991).

Main Massing

A main massing is the greatest concentration of biological diversity within the geographic range of a taxon. This diversity may be measured by taxonomic diversity (e.g., number of subspecies, species, genera), or other aspects of biology such as genetic, phenotypic or behavioral characteristics (Craw 1985). Lesser centers of diversity comprise smaller individual massings.

Main massings are often identified as a particular geographic area, such as a specific continent, land mass, or landscape, but they can only be identified unambiguously as a point at the center of the highest concentration of diversity. In most cases the designation of an area will be sufficient, but when complicated or ambiguous distributions are being analyzed, the geographic center, or center of main massing, may be more appropriate as the point of reference.

Historically, main massings have been used by some biogeographers as a criterion for the center of origin from which the peripheral taxa migrated (Cain 1943, 1944). This is not implied in panbiogeography, where the main massing is used to orient or give polarity to tracks by identifying their geographic relationships to a main massing and to distinguish overlapping tracks associated with different massings. Track polarity is determined by orienting track links from localities more proximate to the main massing to those that are more distant. Overlapping tracks with different main massings will be apparent from the incongruent orientations of the tracks.

Baseline

A baseline is a spatial correlation between a track and a specific geographic or landscape feature (geological or geomorphic) traversed by that track. The baseline allows correlations to be made between distribution patterns and patterns of geology and tectonics, from which hypotheses of historical relationship may be proposed. Tracks sharing the same baseline may be considered biogeographic homologues. This concept of homology is spatial, and contrasts with dispersal biogeography where homology is based on shared means of dispersal, presumed age of origin, presumed center of origin, or presumed association with a particular historical or ecological scenario.

Baseline characters may include prominent features of the landscape (e.g., ocean basin, rift valley, mountain range) or underlying geological structures responsible for geological evolution (e.g., fault system, spreading ridge, suture zone). The baseline does not "prove" any causal association between a track and a geographic character, but it does provide an explicit hypothesis of relationship that may be tested by comparisons with further distributions. The shared defining feature is unique and corresponds to a biogeographic synapomorphy in the same way that a unique derived character represents a synapomorphy in phylogenetic studies of organisms (Craw 1983).

Many groups are widely distributed and may be disjunct across more than one geological/geomorphic feature so that the geographic span of a track alone cannot provide a choice between alternative baselines. In this situation the baseline is chosen for the geographic character that lies within, or closest to, the main massing rather than on its periphery. The underlying assumption of this criterion is that features coinciding with the main massings of a distribution are more closely associated with the evolution of most taxa composing that distribution than are features that coincide with only a few taxa.

By correlating the origin of a distribution with a specific geographic feature and thereby excluding alternative possibilities, the baseline represents a spatiotemporal center for the origin of a track—a panbiogeographic "center of origin." This concept of "origin" contrasts with vicariance and cladistic biogeographical theory that denies any concept of a center of origin (Humphries 1985, Humphries and Parenti 1986).

Case Study: Ratite Birds and the Southern Beeches

Discussion of the disjunct distributions of the ratite birds and the southern beeches (*Nothofagus*) has dominated Southern Hemisphere biogeography for several generations. Neither group has any known means of long-distance dispersal across ocean barriers (the *Nothofagus* nut is dense, heavy, and intolerant of immersion in saltwater; ratite birds are all flightless), although they both occur on widely separated continental and insular land masses. The living ratite birds comprise the ostriches (Struthionidae) of southwestern Eurasia and Africa, the rheas (Rheidae) of South America, the emu and cassowaries (Casuariidae) of Australia and New Guinea, and the kiwis (Apterygidae) of New Zealand (Sibley and Monroe 1990). Extinct ratite families known only from fossils and subfossils are the elephant birds (Aepyornithidae) of Madagascar and the moas of New Zealand (Dinornithidae) (Bledsoe 1988). Large ratite fossils have been reported from the Paleogene in Brazil and Antarctica (Alvarenga 1983, Tambussi et al. 1994). Living *Nothofagus* species are found in southern South America, New Zealand, New Caledonia, New Guinea, eastern Australia, and Tasmania (Linder and Crisp 1996). *Nothofagus* macrofossils and/or fossil pollen are known from all these areas except New Caledonia (Tanai 1986, Hill 1994, Linder and Crisp 1996). The genus is known also from macrofossils and fossil pollens from the Antarctic Peninsula, Transantarctic Mountains, and Ross Sea regions of western Antarctica (Webb and Harwood 1993). All these areas are separated at present by vast expanses of ocean and sea (fig. 1-7).

Ratite birds and southern beeches were problematic groups for dispersal biogeography because their widespread and disjunct distributions contradicted their apparent poor dispersal abilities. Their biogeographic status remained anomalous until the widespread acceptance of continental drift and sea-floor spreading in the 1960s. This geological revolution appeared to solve the biogeographic problem of the ratites and southern beeches by providing a convincing geological mechanism for a credible land bridge. Their disjunct distributions could now be explained by postulating that their ancestors were once members of an ancestral biota confined to the former Gondwana supercontinent (fig. 1-8) because (1) they lacked the necessary means of dispersal to cross the ocean basins, (2) they were "old" or "primitive" taxa (although their fossil history did not necessarily confirm this), and (3) their living representatives are all present on at least some of the land areas that once were part of Gondwana.

The panbiogeographic method can be used to compare and contrast the geographical distributions of these two groups with those sectors of the earth postulated to be involved most intimately with the fragmentation of Gondwana. Fragmentation of the former Gondwana superconti-

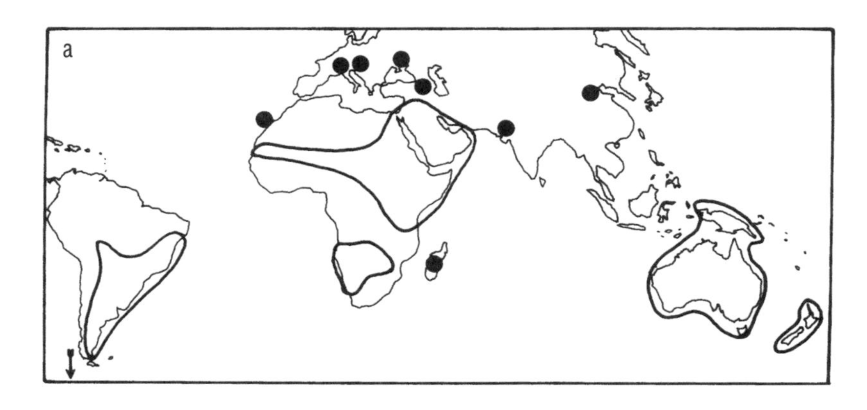

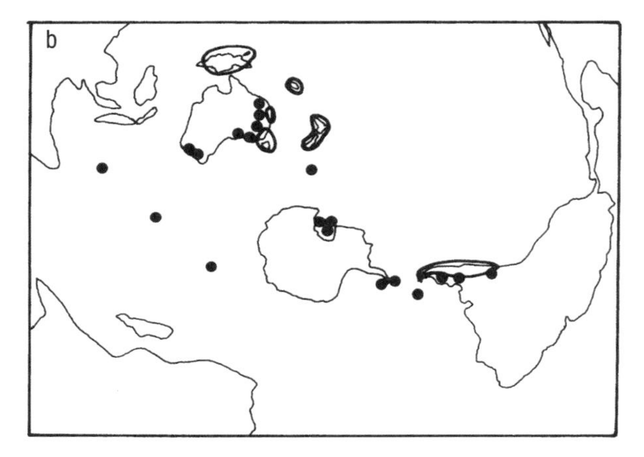

FIGURE 1-7. Generalized distributions of (a) ratite birds (Antarctic ratite fossils indicated by arrow) and (b) southern beeches (*Nothofagus*) (fossil representatives beyond modern range shown as filled circles). Note how both groups are present on major areas associated with the Gondwana supercontinent (distribution data from Craw 1985, Webb and Harwood 1993).

nent requires the formation of two major oceans—the Atlantic (separation of the Americas from Africa and Europe) and the Indian (separation of Africa, Madagascar, India, and Australasia from one another and from Antarctica). If the biogeography of a taxon is associated with the fragmentation of Gondwana, and hence the historical formation of those ocean basins, its track, baseline, and main massings should be centered on or around either one or both of those ocean basins.

The extant and extinct distribution of ratite birds centers around three ocean basins (Atlantic, Indian, and Pacific), with additional fossil records from Europe and Asia (fig. 1-7a). This distribution pattern could be connected by either a trans-Pacific or a trans-Indian and Atlantic Ocean track. Main massings of the ratite birds emphasize the Indian

FIGURE 1-8. Reconstruction of Gondwana and the postulated ancestral distributions of ratite birds and southern beeches (*Nothofagus*) according to the commonly accepted hypothesis of a common Gondwanic origin for both groups (from Page 1990a; reprinted with permission of SIR Publishing). In this model both groups are viewed as having a common biogeographic homology represented by their derivation from West Gondwana.

Ocean with a primary massing of three families in Australasia (Apterygidae, Casuariidae, and Dinornithidae) and a secondary massing of two families on Africa (Struthionidae) and Madagascar (Aepyornithidae). A minimal spanning link between these two massings would cross the Indian Ocean, and then the distribution of the South American rheas would be connected by a minimal spanning link across the Atlantic Ocean to connect with the distribution of the African ostriches (fig. 1-9a). In panbiogeographic terms, the ratite birds can be described as having trans-Atlantic and trans-Indian Ocean tracks and baselines, and their geographical distribution pattern is congruent with a possible Gondwanic origin.

In contrast, the main massings of the southern beeches emphasize the Pacific Ocean, with greatest species diversity running in a line from New Zealand north through New Caledonia to New Guinea. With respect to its South American distribution, *Nothofagus* is described as having a main massing on southwestern South America, as it is confined to Fuego-Patagonia and Chile, largely west of the Andes (Craw 1985).

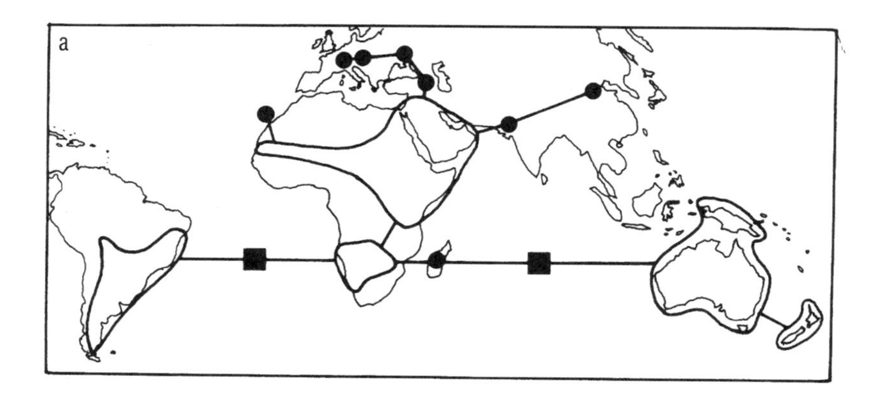

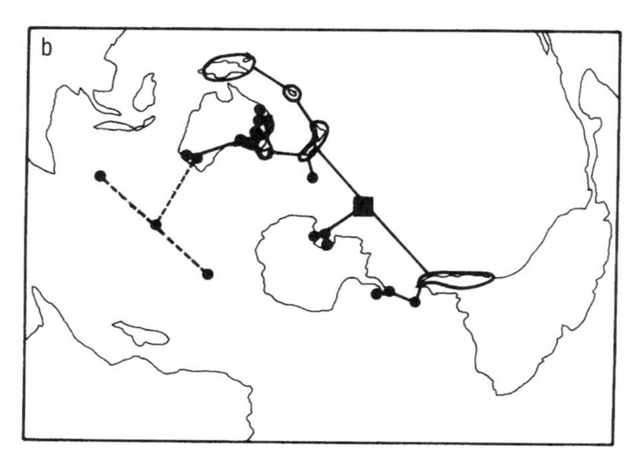

FIGURE 1-9. Tracks, baselines, and main massings of (a) the ratite birds and (b) southern beeches (*Nothofagus*). (Solid squares) baselines; (filled circles) fossil localities beyond modern range. Fossil records are linked to nearest main massings or directly to baseline. In panel b, dashed lines from Western Australia to Indian Ocean connect *Nothofagus*-like pollens. Spatial characteristics do not support a common biogeographic homology for ratites and southern beeches, but suggest different biogeographic and evolutionary histories. Areas of overlap between the two groups are correlated with recognized biogeographic nodes associated with composite geological origins.

Nothofagus is unknown as an autochthonous fossil from Africa and India (Hill 1994), the heart of the Gondwana supercontinent. The track for *Nothofagus* is drawn, therefore, between these two main massings directly over the southern Pacific Ocean (fig. 1-9b) (Croizat 1952, Craw 1985, Humphries et al. 1986, Heads 1990). Southern beeches are described as having a baseline on the South Pacific Ocean. They are a non-Gondwanic group in the sense that their track, baseline, and main massings are not spatially congruent with lands bordering the Indian or Atlantic Oceans.

An alternative explanation is that the distribution of the southern beeches represents a trace of a former west Gondwanic origin along the coast of Antarctica (e.g., Hill 1994). This is the Weddellian biogeographic province hypothesis of Case (1989), who postulated that the region extending from southern South America along the Antarctic Peninsula through West Antarctica to southeastern Australasia was a major center of evolutionary differentiation for *Nothofagus* and for numerous other animal and plant groups. While acknowledging that *Nothofagus* has had a long history in this region, it must be emphasized that the biogeography of the southern beeches cannot be divorced from that of related taxa in the fagalean alliance. For instance, the northern beeches *Fagus* also display a trans-Pacific track, albeit in the North Pacific (fig. 1-10a). They are widely distributed across Eurasia and North America (living and fossil), with a main massing of species in East Asia (China) that exhibit phylogenetic links to eastern North American species (Melville 1982). Distributions of the southern and northern beeches are part of a classic Pacific Rim biogeographic pattern (e.g., Camp 1948), with an extension into Europe and eastern North America (Melville 1982).

As with many groups of biogeographical interest, the phylogenetic relationships of *Nothofagus* are widely debated, and various associations have been considered with other Fagaceae and Betulaceae (e.g., Jones 1986). Geographic distributions of many of these putative relatives of *Nothofagus* support the recognition of the genus as a Pacific, rather than a Gondwanic, group (fig. 1-10). *Nothofagus* is related to Balanopaceae of the southwest Pacific, the transcentral Pacific *Lithocarpus* (western North America and eastern Asia), the transcentral Pacific generic pair *Chrysolepsis* (California) and *Castanopsis* (Southeast Asia), and the transtropical Pacific trigonobalanoid group of genera (Colombia, Southeast Asia) (Forman 1966, Schuster 1976, Melville 1982, Nixon and Crepet 1989, Heads 1990). Distributions of these taxa (except Balanopaceae, which partially overlap with *Nothofagus* in New Caledonia) represent trans-Pacific extensions of the Southern Pacific track of *Nothofagus*, with main massings centered on the Pacific Rim.

Distribution patterns of the ratite birds and the southern beeches are in no way biogeographically homologous or congruent, nor were they both once members of a widespread, ancestral Gondwana biota as is often suggested in the biogeographical literature (e.g., Humphries 1981, Patterson 1981). *Nothofagus* is a member of a non-Gondwanan trans-Pacific fagalean alliance; the ratites are a Gondwanic group centered on the South Atlantic and Indian Ocean basins. In the Southern Hemisphere these groups are geographically sympatric only in southwestern South America and in eastern Australasia where their tracks intersect.

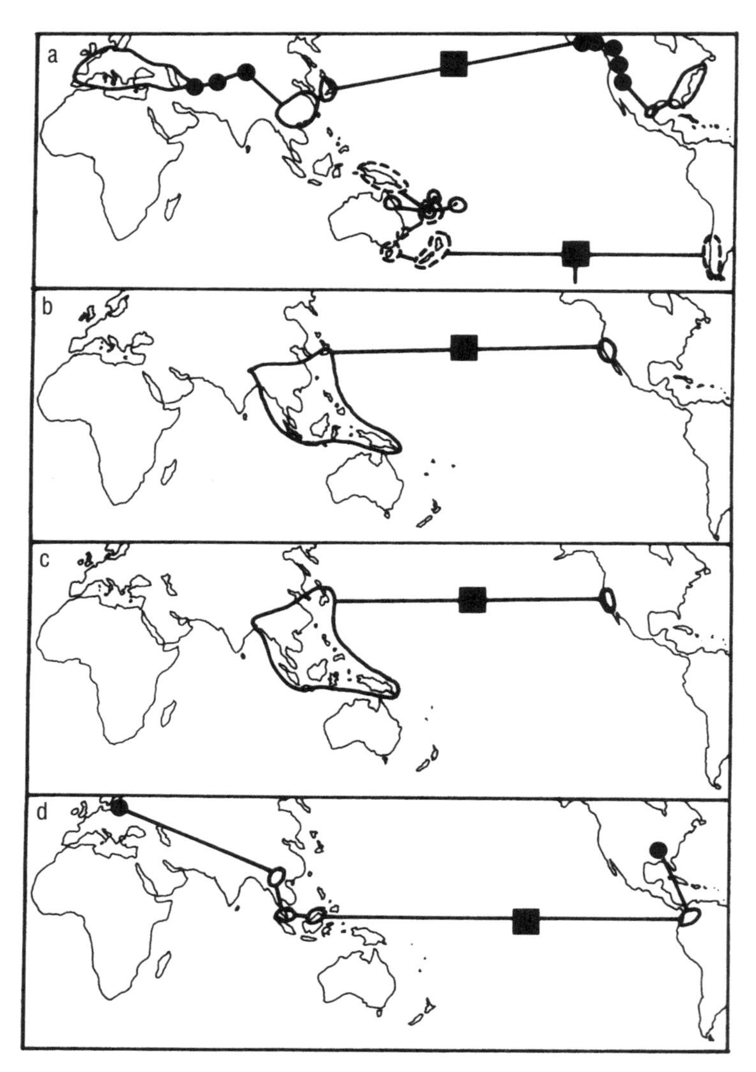

FIGURE 1-10. Pacific Rim distribution pattern of the southern beeches (*Nothofagus*) and related taxa. (Solid lines) extant distributions; (filled circles) fossil records beyond modern range; (filled squares) baselines. (a) Balanopaceae (solid line S.W. Pacific), *Nothofagus* (broken line), *Fagus* (solid line, Northern Hemisphere), (b) *Lithocarpus*, (c) *Castanopsis*, *Chrysolepsis*, and (d) trigonobalanoid genera. In panel d, the track link for the European fossil record is oriented to the Southeast Asian massing (two genera), and that for the North American fossil is oriented to the nearest neighbor in Columbia. These Pacific baselines support a non-Gondwanan origin for the southern beeches (after Schuster 1976, Melville 1982, Zhengyi 1983, Craw and Page 1988, Nixon and Crepet 1989).

These areas can be described as nodes, and they were recognized by Croizat (1952) as the Magellanian and Polynesian gates (prominent nodes), respectively. These nodes can be interpreted as zones where fragments of two or more ancestral biotic and geologic worlds have fused into one modern geography. This interpretation receives considerable support from both biogeographical studies and geological history (Craw 1982, 1988).

1.4 Conclusions

Panbiogeography is a constructive approach to biogeographical study that addresses current and historical debates over pattern and process. It attempts to resolve the dispersal–vicariance opposition that has plagued the discipline since the nineteenth century. This resolution takes the form of a vicariant form-making model which recognizes both dispersal and vicariance as important processes by which organisms achieve their geographic distributions.

Panbiogeographic method emphasizes the analysis of raw locality and broader distribution data for taxa. This is a different approach to the a priori assignment of taxa to already defined biogeographic elements, regions, and areas of endemism that, to date, has been the most prevalent mode of analysis in historical biogeography. Application of this method can lead to the proposal of novel hypotheses concerning the biogeography of many different groups of animals and plants. But most of all, the panbiogeographic project reaffirms the importance of the geographical context to any understanding of the history of life. In biogeography, as in the real estate business, there is only location, location, location.

2

Life as a Geological Layer

Panbiogeography and the Earth Sciences

Biogeography is a disciplinary site where the earth sciences intersect dramatically with some of the most fundamental tenets of evolutionary biology. It is a complex discipline combining knowledge of geographical distribution and evolutionary principles with geological and geomorphological settings. If speciation and the origin of higher clades, dispersal, vicariance, and extinction events have occurred over the same temporal and spatial scales as earth history events, then the historical patterns of organic distribution may never be erased completely through time and across space. Biogeographical explanations require an understanding of how geographic distribution is produced and of how the spatial fabric and texture of biotic communities derive from this historical legacy of hundreds of millions of years of climatic and tectonic change and biotic interactions (Brown 1986). The challenge of making practical and theoretical sense of this historical legacy left in organic distribution is leading to the formation of a distinctly different biology of space, time, and form.

Geographical distributions can be constrained historically by geological events and processes, though they may indeed also be in ecological equilibrium with present conditions. The geographic distribution of life is controlled by the configuration of land, sea, and other topographic features and climate. All of these factors are governed by the motion of the tectonic plates that compose earth's outer shell, its lithosphere, to a significant extent. Biogeographical consequences of plate motions and interactions are enormous. The rearrangement of continental landmasses and island arcs and the opening and closing of sea and ocean basins initiated by these motions and interactions have profoundly affected the distribution and history of organisms (McKenna 1983, Hallam 1994).

Panbiogeographic studies suggest that global and regional patterns of tracks and nodes may have their origin in a Mesozoic to early Tertiary geography. These tracks involve taxa at various levels of the taxonomic hierarchy, including species, genera, and families. Biogeographical pattern analysis combined with fossil evidence suggests that Mesozoic and early Tertiary events were a major influence on the dispersal and vicariance of ancestral biota leading to present-day global and regional distribution patterns. Pleistocene glaciations resulted in sometimes drastic local and regional disruptions to distributions, but pre-Pleistocene patterns were not obliterated (Croizat 1961).

2.1 Pleistocene or Earlier?

Quaternary history began to dominate biogeographical analysis during the 1950s because it was thought that Pleistocene glaciations had changed the whole underlying patterns of plant and animal distribution and had had profound effects on evolution (Deevy 1949, Holdhaus 1954, Darlington 1957, De Lattin 1967). Numerous authors have proposed large-scale extinctions and extensive speciation mediated by Pleistocene glacial cycles, but the relationships between modern distributions and postulated Quaternary climate patterns are often problematic. Rosen (1978), for example, comments on the lack of congruence between paleogeographic maps showing a hot, dry-climate flora in central and northeastern Mexico, southern and western Texas and the mountain states at the glacial maximum, whereas a cooler and wetter climate is required for a forest connection across north-eastern Mexico and southern Texas.

Recently, more attention has been paid to the Tertiary background of faunal and floral developments worldwide. The conclusion of Coope (1970, 1978) and Matthews (1977) from the entomological fossil record is that there is no evidence of morphological evolution during the last half million years and that Pleistocene extinctions have played a very minor part in the evolution of present-day insect faunas. Coope (1994) documented and compared distributions of fossil and living beetles (Coleoptera) in Europe and found that species remained constant in both morphology and environmental requirements throughout the Quaternary. Changes to geographic range could be related to the last glacial, but Coope (1994) suggested that similar changes in species range accompanied each of the major climatic oscillations. The numerous large-scale climatic changes contributed to local, often catastrophic, extinctions, but global extinctions were found to be remarkably rare.

Molecular clock studies on amphibians, particularly for Neotropical and Australian groups (e.g., Maxson and Heyer 1982, Barendse 1984, Maxson and Roberts 1984) have shown that in most cases vicariant events responsible for subdivisions of monophyletic clades into species and species groups occurred during the Tertiary rather than during the Quaternary era. Similar findings have been made for European amphibians (Szymura 1983). A study of the systematics and distributions of the Palearctic amphibians led to the conclusion that the amphiboreal gaps in their distributions may have appeared before the Pleistocene events (Borkin 1986). The two major groups of Holarctic hylid frogs may have diverged as long ago as the early Eocene (50 MYBP), and certainly no later than the mid-Oligocene (27 MYBP) (Hedges 1986). Many basic statements of the glaciation hypothesis do not appear to be supported by recent data. Information from cladistic systematics, geographic distribution, paleontology, and molecular studies suggests that disjunctions often attributed to the effects of the Pleistocene glaciations are the result of earlier divergence (Tangelder 1988, Klicka and Zink 1997).

Similar conclusions were reached by Allen (1990), who questioned the hypothesis that endemism in the interior highlands of North America is the result of events associated with Pleistocene glaciations, especially since the region under study has been an above-water landmass since the Pennsylvanian era (c. 320 MYBP). Major evolutionary patterns of the early Tertiary can still be traced in the biogeography and phylogenetic relationships of extant North American pines. Pleistocene climatic changes do not appear to have completely reshuffled the distribution of *Pinus*, and Tertiary distribution patterns and the traces of evolutionary events dating from the Eocene period may have been maintained into the present (Millar 1993). Molecular phylogenies and geographical distributions of western North American aridland rodent faunas indicate that mid- to late Pleistocene events had little influence on lineage divergence. Divergence and current distributions appeared instead to be effects of the late Tertiary development of the western North American cordillera (Riddle 1995). Scenarios for speciation of northwest Pacific trout and salmon often involve Pleistocene glacial advances and retreats as the pumps for isolation, diversification, or behavioral modification, but Stearley (1992) argues that the accumulating fossil evidence indicates that most of the differentiation leading to the origins of the extant species occurred before the ice ages, in the Pliocene and Miocene.

High species diversity in tropical forests has been explained as originating through the contraction of rainforests into vicariant refugia during Pleistocene glacial advances. Isolated geographically by these contractions, rainforest animals and plants underwent speciation (Haffer 1969). As an alternative, Cracraft (1985) proposed that in many of these areas in South America endemism is considerably older than the Pleistocene. Bush (1994) has shown that pre-Quaternary vicariance events appear to have established the major regional divisions of species complexes in Amazonian South America. Bush's study suggests that the central Amazonian forests did not fragment during glacial times. Molecular phylogenetic studies of Australian and South American rainforest animals suggest that most divergence and speciation events occurred before the late Pleistocene (Brower 1994, Fjeldsa 1994, Joseph et al. 1995).

Mesozoic tectonics have been recognized as a factor in the origin of transoceanic disjunct distributions, and many biogeographical studies suggest that the current ranges of taxa are congruent with hypothesized historical continental alignments of that era. For instance, Carlton and Cox (1990) postulated that the trans-North Atlantic track linking the distributions of amauropsine ant beetles (Coleoptera: Pselaphidae) in eastern North America and Europe represented the trace of an ancestral lineage. An amauropsine ancestor was distributed along the Hercynian, Appalachian, and Marathon-Oachita Thrust belts in the middle Jurassic, a time when Europe and eastern North America were contiguous (fig. 2-1a). By the lower Cretaceous, initial opening of the North Atlantic had isolated descendant lineages in Europe (*Amaurops* and *Paramaurops*) and North America (*Arianops*) (fig. 2-1b). In late Cretaceous time, development of the Mississippi embayment fragmented the ancestral American *Arianops* population into Appalachian and Interior Highland lineages (fig. 2-1c). Similarly, the disjunction between South Africa and Australia for chironomid flies confined to pre-Jurassic rocks on each continent has been related to fragmentation of an ancestral range by seafloor spreading of the Indian Ocean (Cranston et al. 1987).

Though there was certainly an influence on diversity (speciation and extinction) and distributions (alternating north and south range movement) due to the Pleistocene glaciations and associated climatic changes, the histories of many living animal and plant groups are distinctly rooted in the late Mesozoic and early to mid-Tertiary. Climatic variation would most likely have occurred during the Quaternary over preestablished patterns of distribution promoting only expansion and retraction of biota around centers of endemism that originated earlier. Recent events may have influenced distribution patterns, but this is not justification for ignoring the impacts of earlier events. The main patterns of organic distribution, even in regions directly influenced by glaciation, correspond to early to mid-Tertiary or Mesozoic events (Tangelder 1988, Amorin 1991).

2.2 Fossil Evidence

Distribution records, whether living or fossil, provide the database of panbiogeography, but some critics have asserted that fossils are over-

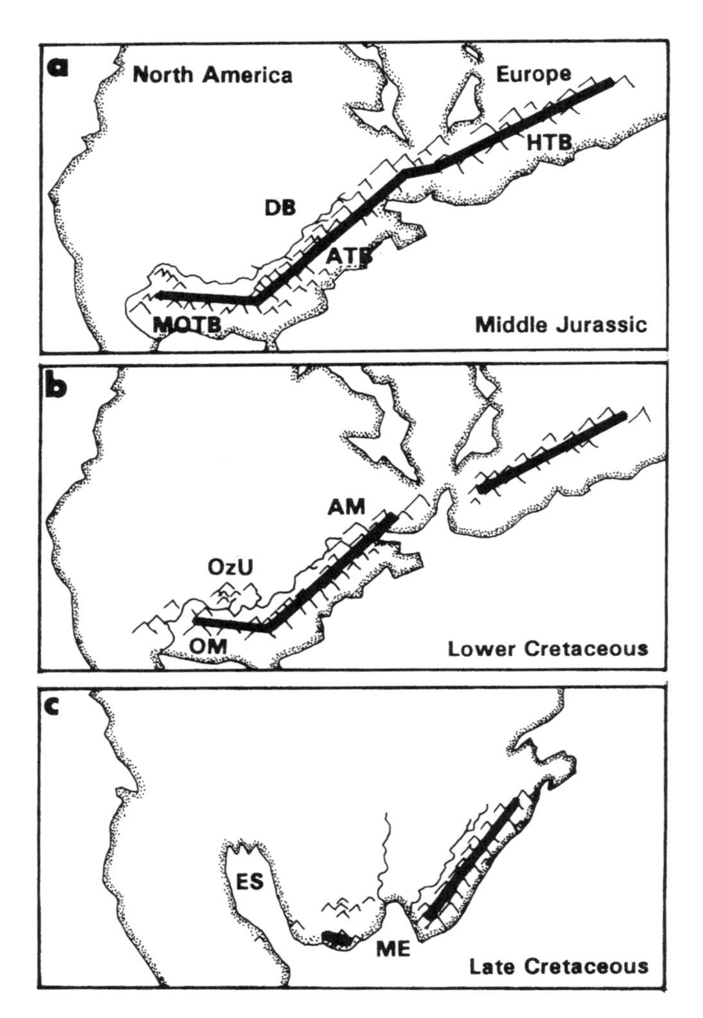

FIGURE 2-1. A correlation between current distribution and hypothesized Mesozoic continental alignments as illustrated by pselaphid beetle tracks (tribe Amauropsini). (a) Middle Jurassic: ancestral track continuous between North America and Europe along Hercynian Thrust Belt (HTB), Appalachian Thrust Belt (ATB), Marathon-Ouachita Thrust Belt (MOTB), and Ozark Uplift (OzU); (b) lower Cretaceous: ancestral track fragmented by initial opening of North Atlantic with continuous North American track retained along the Appalachian Mountains (AM), and Ouachita Mountains (OM); (c) late Cretaceous: North American ancestral *Arianops* track fragmented by development of Mississippi Embayment (ME), and epicontinental seaway (ES) (from Carlton and Cox 1990; fig. 2; reprinted with permission of C. Carlton).

looked or denied in this research program (Cox and Moore 1993, George 1993). Living records do constitute the main source of information on distribution patterns, as they are far more extensive than fossil records of the same or related taxa. Fossil records provide additional localities and taxa for analysis, rather than proof of absolute place and time of origin. The fossil record is incomplete, and earliest occurrences of a particular group may yield only upper bounds on divergence times. These upper bounds can significantly underestimate true divergence times (Marshall 1990). When the fossil record is fragmentary or poor, as in the case of mammals, actual rather than minimal divergence times may be very different (Easteal et al. 1995). First and last occurrences of fossils will inadequately approximate times of origination and extinction (Holland 1995).

Fossils are recognized in panbiogeography as representing the minimal age of a group, which is not necessarily equivalent to the time of phylogenetic origin of a group. Age of origin is at least slightly older than the oldest age of fossilization (Croizat 1964). This important distinction is often overlooked in biogeography, where the trend has been to produce biogeographical scenarios based on the place and geological time of first appearance of a group in the fossil record. But such an approach is vulnerable to the trend toward discovery of ever older fossils referable to extant groups. For example, recent fossil discoveries have extended by many millions of years the known geological time range of hagfish and pseudoscorpions. Newly discovered middle Devonian pseudoscorpion fossils have extended the known range of this taxon from 35 million years (mid-Tertiary) to 380 million years (mid-Paleozoic) (Shear et al., 1989), while the fossil hagfish is about 300 million years old and is the first reported fossil for this group (Bardack 1991).

Many other extant groups are also ancient. Family differentiation of modern freshwater Mollusca had taken place by the late Paleozoic, for many well-preserved fossils from this period are readily assignable to living families or even subfamilies, and some extant genera can be recognized by early Cretaceous times (Taylor 1988). Mid-Cretaceous fossil mushrooms bear a strong resemblance to existing genera (Hibbert et al. 1995), and the living Queensland lungfish has lived in Australia at least since the early Cretaceous (Long 1992), suggesting that certain modern morphologies may be of ancient origin. Fossil taxa closely related to several groups of extant birds are known from late Cretaceous rocks. Phylogenetic inferences strongly suggest that at least half a dozen lineages of modern birds were differentiated before the end of the Mesozoic (Chiappe 1995). Further fossil evidence indicates that many other living groups had appeared in modern form at the family level by the Mesozoic, and sometimes even earlier (table 2-1).

Known fossils show the long time range of numerous living genera and species. Paleocene fossil fruits from North America are as-

TABLE 2-1. Earliest fossil records for some extant animal and plant families (fossil data from Benton 1993)

Taxonomic group	Geological period
Ferns and Fern Allies	
Lycopodiaceae	mid-Devonian
Equisetaceae	Permian/Triassic
Selaginellaceae	late Devonian
Marrattiaceae	Carboniferous
Osmundaceae	late Permian
Gleicheniaceae	Permian
Matoniaceae	Triassic
Dipteridaceae	Triassic
Polypodiaceae	Triassic
Dicksoniaceae	Triassic
Schizaceae	Jurassic
Cyatheaceae	Cretaceous
Loxosomaceae	Jurassic
Ophioglossaceae	early Tertiary
Marsileaceae	Cretaceous
Seed Plants	
Gnetaceae	Triassic
Podocarpaceae	Triassic
Araucariaceae	Triassic
Pinaceae	Triassic
Taxodiaceae	Jurassic
Cupressaceae	early Tertiary
Ginkgoaceae	Triassic
Fagaceae	Upper Cretaceous
Hamamelidaceae	Lower Cretaceous
Magnoliaceae	Cretaceous
Nymphaceae	Cretaceous
Platanaceae	Upper Cretaceous
Insects	
Springtails	
Entomobryidae	Triassic
Isotomidae	Triassic
Mayflies	
Ephemerellidae	Triassic
Leptophlebidae	Triassic
Siphlonuridae	Jurassic
Dragonflies	
Aeshnidae	Jurassic
Gomphidae	Jurassic
Petaluridae	Jurassic
Earwigs	
Labiidae	Jurassic
Crickets	
Prophalangopsidae	Triassic
Stoneflies	
Eusthenidae	Permian

TABLE 2-1. *(continued)*

Taxonomic group	Geological period
Insects	
Bugs	
Corixidae	Jurassic
Notonectidae	Triassic
Miridae	Jurassic
Beetles	
Byrrhidae	Jurassic
Carabidae	Jurassic
Cerambycidae	Early Cretaceous
Chrysomelidae	Jurassic
Cupedidae	Triassic
Elateridae	Jurassic
Hydraenidae	Jurassic
Scarabaeidae	Jurassic
Silphidae	Jurassic
Flies	
Chironomidae	Jurassic
Mycetophilidae	Jurassic
Tipulidae	Triassic
Caddisflies	
Philopotamidae	Jurassic
Moths	
Micropterygidae	Jurassic
Nepticulidae	Jurassic
Vertebrates	
Fish	
Amiidae	Jurassic
Lepisosteidae	Cretaceous
Polypterigidae	Cretaceous
Squalidae	Jurassic
Frogs	
Discoglossidae	Jurassic
Leptodactylidae	Cretaceous
Pipidae	Cretaceous
Turtles	
Chelyridae	Cretaceous
Pelomedusidae	Cretaceous
Trionychidae	Cretaceous
Lizards	
Gekkonidae	Cretaceous
Iguanidae	Cretaceous
Scincidae	Cretaceous
Varanidae	Cretaceous
Xenosauridae	Cretaceous
Mammals	
Ornithorhyncidae	Cretaceous

signable to the extant freshwater angiosperm genus *Ceratophyllum*, and mid-Eocene and upper Miocene fossils could be referred to existing species of that genus (Herenden et al. 1990). Fossil leaves of four modern genera of Euphorbiaceae have been documented from the middle and upper Eocene of Japan (Tanai 1990). Study of a European Eocene assemblage of liverworts in the order Jungermaniales showed that most of the fossils could be assigned to modern genera (Benton 1993). Many insect genera are of great age as shown by the modern fauna found in early Tertiary Baltic amber (Larsson 1978). An Eocene fossil final instar larval marchfly (Diptera: Bibionidae: *Dilophus*) from New Zealand is close to a common extant species, and Mesozoic bibionid adults closely resembled modern species (Harris 1983). Three genera of decapod crustaceans and several asteroid echinoderms of Eocene age are all similar to living forms (Blake and Zinmeister 1988, Feldmann and Wilson 1988), and four extant salamander genera are known from skeletal fossils as old as Paleocene time (Naylor and Fox 1993).

There is some suggestive evidence that it is not only the lineages of modern taxa that may be ancient. Extant faunal assemblages and botanical associations may have long roots extending back beyond the Pleistocene to the late-to-mid–Tertiary period, and sometimes even into the late Mesozoic. A lowermost mid-Cretaceous insect fauna from Australia, comprising some 70 taxa, is an essentially contemporary fauna with numerous taxa virtually indistinguishable at the generic level from those of today. This fauna is similar both taxonomically and ecologically to present-day insect communities in the same area, indicating stasis since the late Mesozoic (Jell and Duncan 1986). Zones of tectonic stability in the central North Island of New Zealand have fossil landform surfaces that may represent a well-preserved terrestrial Miocene landscape and associated old plant cover, which has survived the vicissitudes of Pliocene marine transgressions and Pleistocene glaciations (Rogers 1989). Marine communities may also exhibit considerable stasis through time. For example, Petuch (1981) has described a relict fauna of 45 species of caenogastropod Mollusca from shallow water along the northern Colombian and Venezuelan coasts of South America. These living gastropods were previously known only as fossils from Miocene and Pliocene strata, and they appear to have survived both Plio-Pleistocene sea level and temperature changes.

Although many taxa have been assumed to have a recent biogeographical history, a late Mesozoic-early Tertiary origin for a broad range of modern animal and plant taxa is supported by the fossil record. Even contemporary species and genera may have fossil records compatible with late Mesozoic and early-to-mid–Tertiary origins, and living families can have records supporting a phylogenetic origin by at least early Mesozoic times. Fossil evidence read as the minimal age of fossilization supports a Mesozoic and early Tertiary origin for the present biogeographical patterns of much modern life.

Case Study: Fossils and the Biogeographical Evolution of Angiosperms

Problems exposed in biogeographical explanations based on the place and time of first occurrences of a group in the fossil record can be exemplified by recent proposals concerning the origin and evolution of the flowering plants. Cox and Moore (1993) illustrate this assumption with their timetable for an early Cretaceous origin and dispersal of angiosperm plants. This post-Jurassic scenario rests primarily on the apparent absence of undisputed angiosperm fossils before Barremian time (early Cretaceous) and the premise that the earliest origin of angiospermy dates from about this time. In contrast, phylogenetic treatments (Crane 1985, Doyle and Donoghue 1986, 1987) suggest that angiosperms are the same age as the closely related Bennettiales and Gnetales, which have fossil records extending back into the Triassic. A pre-Jurassic origin for angiosperms is supported by molecular clock studies that have dated the divergence between dicotyledonous and monocotyledonous flowering plants sometime in the Triassic or earlier (Clegg 1989, Martin et al. 1989, Martin and Dowd 1991, Savard et al. 1994). Late Triassic and upper Jurassic angiosperm pollen have been described (Cornet 1989, Cornet and Habib 1992), and a late Triassic angiosperm macrofossil has been reported (Cornet 1986). These discoveries, and other data discussed by Heads (1984) clearly require a reassessment of theories postulating a first effective radiation of angiosperms in early Cretaceous times.

The biogeography and systematics of the cotton plants (*Gossypium*, Malvaceae) provide a further illustration of the problems inherent in establishing an upper boundary on divergence time based on first fossil occurrences. The group is of evolutionary, as well as biogeographic, interest because it includes some tetraploid species, and recent taxonomic revision has followed upon the application of molecular techniques accompanied by molecular clock estimates for the age of divergence from closest relatives.

Cotton plants range through much of the New World, Africa, Asia, and Australia (Hutchinson et al. 1947). The absence of cottons from southwest South America and the presence of wild cottons in northwestern and central Australia is consistent with Atlantic and Indian Ocean tracks and baselines (fig. 2-2). Although cottons are highly mobile and can apparently disperse virtually anywhere, wild cottons exhibit vicariant distribution patterns wherever they occur, suggesting that immobilism dominated their differentiation. These baselines and vicariant distributions imply that the present transcontinental range re-

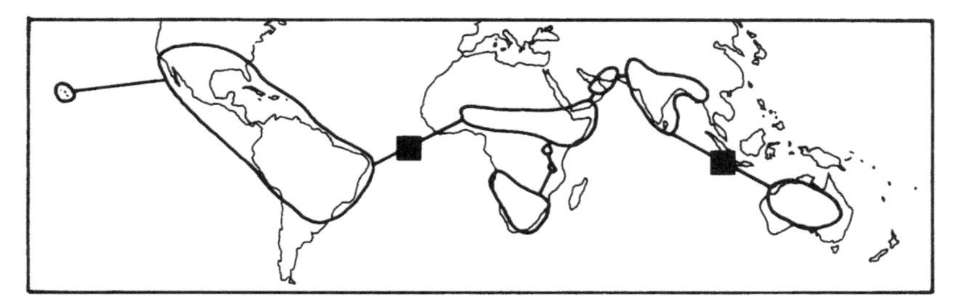

FIGURE 2-2. Generalized distribution of the cotton genus *Gossypium*. Track baselines (filled squares) in the Atlantic and Indian Ocean basins may suggest fragmentation of a widespread Gondwanic ancestral range (after Hutchinson et al. 1947, Saunders 1961, J. F. Wendel, personal communication 1995).

sulted from the fragmentation of a formerly contiguous ancestral range across Gondwana through the formation of the Indian and Atlantic Oceans (Hutchinson et al. 1947, Croizat 1958, 1964; Fryxell 1979).

A recent molecular phylogeny for *Gossypium* based on cladistic analysis of chloroplast DNA restriction site variation identified three main lineages: (1) New World tetraploids and African–Asian (Indian) diploids, (2) diploid New World–African cottons, and (3) diploid Australian cottons. A biogeographic analysis matched taxon/area cladograms for congruence with the sequence of continental fragmentation according to the "Terra Mobilis" computer program. Congruence between the area cladogram and this program led to an initial hypothesis that the divergence between African and Australian cottons had occurred 130–120 MYBP, the most recent period at which *Gossypium* could have had a continuous ancestral distribution across Gondwana. This divergence date was considered dubious because it is twice as ancient as the oldest recorded malvaceous (similar to that of the family to which the cottons belong) fossil pollen, which was considered evidence of an Eocene or younger origin for the family Malvaceae (Wendel and Albert 1992).

To solve the apparent contradiction between fossil evidence and the congruent biogeographic, systematic, and geological evidence, Wendel and Albert (1992) introduced a reassessment of cotton divergence times using selected plant taxa and fossil records, which dated the initial dichotomy in the upper to mid-Oligocene (i.e., 24–32 MYBP). The African and Australian continents were separated by wide ocean gaps by this period, so the initial disjunction between the African and Australian cotton lineages, as well as all later intercontinental divergences, were explained through long distance dispersal rather than vicariance. This revised timetable privileges the oldest known fossil record over congruence between biogeographical, geological, and systematic hypotheses.

But molecular clock estimates standardized by a minimal age of fossilization could considerably underestimate divergence ages (Springer 1995). For example, while paleontological evidence indicates that mouse and rat lineages diverged between 8 and 12 MYBP, independent immunological and DNA/DNA hybridization studies have estimated this divergence at between 17 and 35 MYBP. Divergence of mice and rats may be much older than suggested by a literal interpretation of the fossil record (Janke et al. 1994). Hedges (1996) confronted this problem for the diversification of mammalian orders by using the fossil divergence time between diapsid and synapsid reptiles (310 million years) as an external calibration point resulting in a Mesozoic, rather than Cenozoic, molecular divergence time estimate for avian and mammalian orders.

2.3 Geological and Biogeographical Correlation

Patterns of biogeographical distributions do not necessarily correlate with exposures of particular strata or rock types. Instead, biogeographical distributions may be congruent with zones of tectonic activity (e.g., plate and terrane margins, fracture zones, and belts of granite), or past coastlines and their environs so that much of present-day biology is the product of old, relic communities left stranded on more recent plains, hills, mountains, and landmasses, all by normal geological processes such as accretion, collision, erosion, orogenic uplift, volcanic eruption, and eustatic sea level changes.

These interactions between ecology, geology, and history are not confined to one restricted temporal or spatial scale. Significant effects on animal and plant distributions have been postulated for Pleistocene river catchments (fig. 2-3a) and ancient coastlines (fig. 2-3b). Major tectonic features may be associated with the distributional limits of even widespread taxa. The East African tectonic rift, for example, lies at the boundary between two vicariant bird distributions (fig. 2-3c). On an even broader spatial scale, the limits of marine fish distributions may be correlated with major plate tectonic boundaries (fig. 2-3d).

An evolutionary interrelationship between ecology, earth history, and geography was noted by Taylor (1966) in a biogeographical study of living and fossil freshwater Mollusca. Tectonically active areas with changing drainage patterns and habitats were found to be rich in local forms that were restricted stratigraphically and geographically. Tectonic activity was postulated to promote molluscan differentiation by changing environments and separating or joining habitats. Both the formation of new communities from previously separated ancestral biota and the fragmentation of ancestral populations leading to speciation are associated with geological processes governing stream confluence, stream capture, and fragmentation of habitat (Taylor and Bright 1988). Taylor

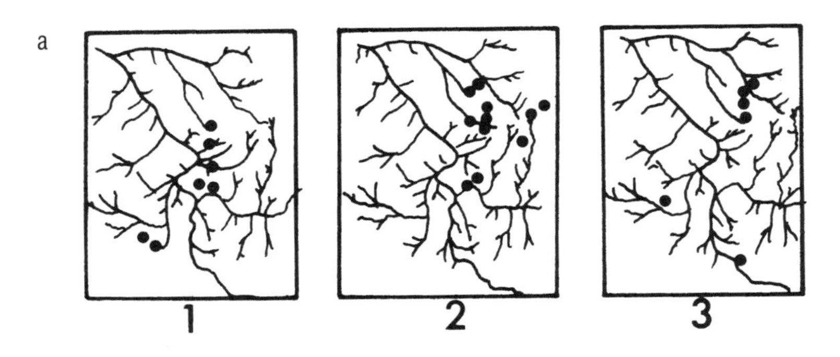

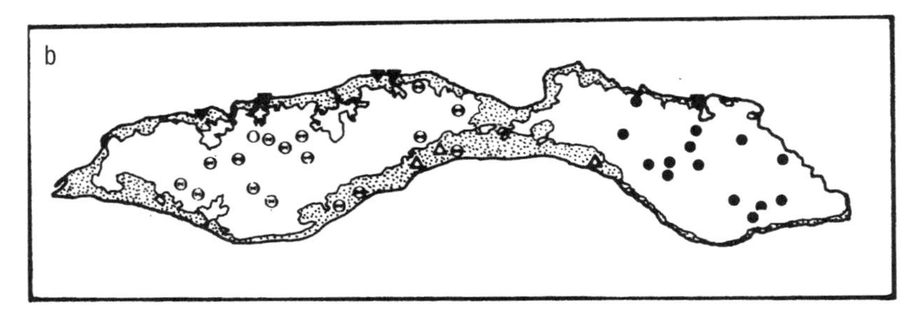

FIGURE 2-3. Various spatiotemporal scales of geomorphological and tectonic correlations with biological distributions. (a) Pleistocene fossil drainage systems and plant species. Stuessy (1990) correlated the distributions of some plant species in Ohio with glacial-age rivers and suggested that this biogeographic pattern arose from migration by the plants into the area along Pleistocene drainage patterns. 1, *Magnolia tripetala*; 2, *Phlox stolonifera*; 3, *Rhododendron maximum* (after Stuessy 1990). (b) Pleistocene sea-level changes and amphipod crustaceans. Stock (1977) correlated the distributions of endemic groundwater amphipod species distributions on Bonaire Island (Netherlands Antilles) with different regressive sea levels (unshaded = two Miocene areas, with Pleistocene/Holocene terraces to the north and south). (c) The African rift zone and continental distribution of starling species. Craig (1985) suggested that each of four species (represented by different symbols) could be characterized by their positions to the east, west, and south of the rift (from Craig 1985; reprinted with permission of *Ostrich*). (d) Marine fish species and the Pacific plate boundary (from Springer 1982; reprinted with permission of Smithsonian Institution Press).

(1960) identified this kind of correlation as an example of a panbiogeographic approach in an analysis of fossil and living clam distributions in relation to ancient and modern drainage patterns. A similar example for a much earlier time period is the observation that Ordovician radiations of trilobites, brachiopods, bivalves, gastropods, and monoplacophorans were far more diverse in and near foreland basins than they were in areas removed from orogenic activity. This suggests an associa-

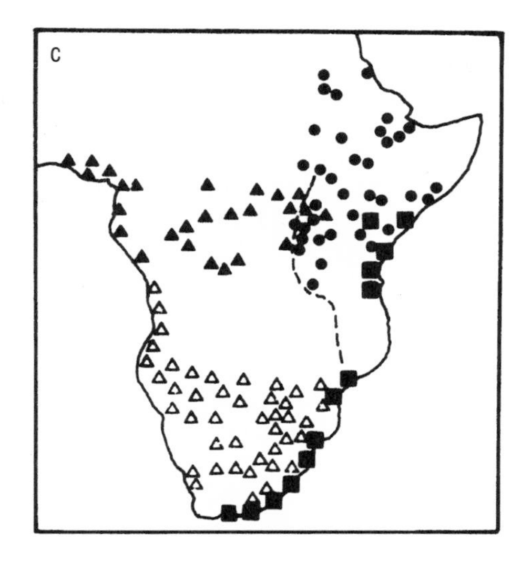

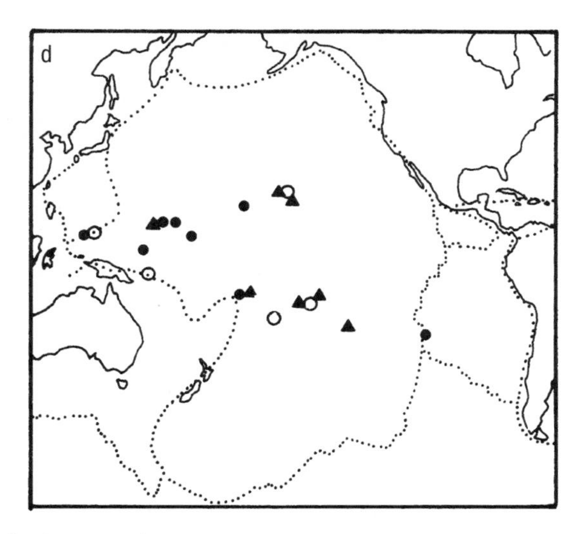

FIGURE 2-3. *(continued)*

tion of orogeny with biological diversification in that geological time (Miller and Mao 1995).

Tectonic and other geological activity tends to result in more irregular land and seafloor surfaces, and these irregularities promote biological diversification. Though radically different in detail, the distributions and ecologies of organisms as diverse as snails, birds, fishes, and trees may evolve through the same relationship with changing geographic, tectonic, and geological processes, whether by uplift, subduction, sedi-

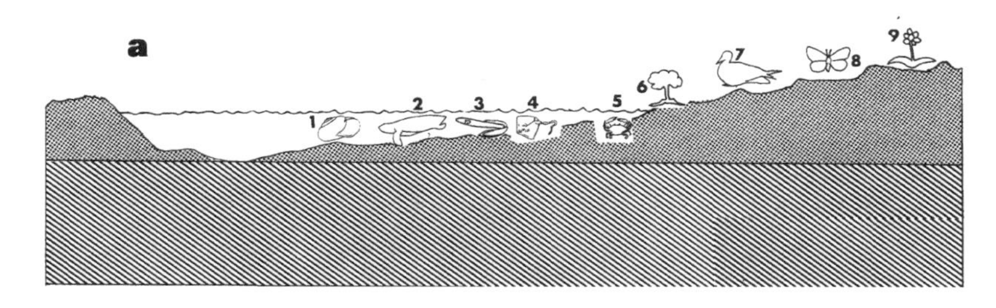

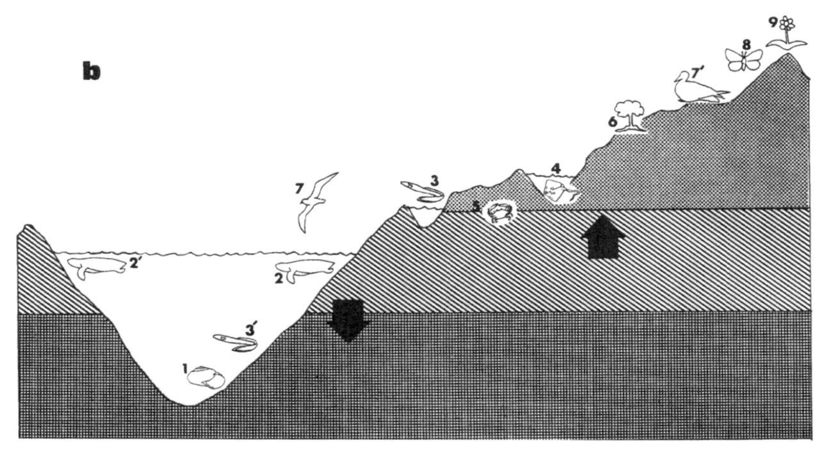

FIGURE 2-4. A model illustrating common biogeographical and geological factors in the evolution of ecological communities. (a) Ancestral organisms distributed in a region of low relief such as a coastal plain next to a shallow-water estuarine habitat in the Mesozoic or early Tertiary. Ancestral organisms: 1, marine mollusc; 2, marine turtle; 3, coastal eel; 4, marine stingray; 5, interstitial shore Crustacea; 6, mangrove; 7, shore-nesting sea-bird; 8, coastal moth; 9, coastal herb. (b) Tertiary evolution of the landscape into a major deep-water sea basin with adjacent upland and montane habitats. Descendent organisms: 1, marine mollusc now deep-water seafloor; 2, feeding and breeding habitats of marine turtle now geographically distant; 3, coastal eel with marine breeding and freshwater adult development; 4, stingray now freshwater; 5, interstitial shore Crustacea now subterranean freshwater; 6, mangrove now terrestrial upland montane; 7, sea-bird marine feeding, montane nesting; 8, 9, coastal moth and herb now sub-Alpine/Alpine.

mentation, or redeposition (fig. 2-4). Through these geological processes, older life may become established on new landscapes. Young mountains may be worn down, and the organisms once located on young sediments may now rest upon very old metamorphosed rocks. An old peneplain may be subjected to marine transgression and then uplifted to provide a new surface onto which nearby organisms may migrate and establish new ecological communities. Similarly, old landscapes may be covered with younger sediments or volcanic debris. Despite all these geological upheavals, organisms may survive in the general vicinity by moving between these changing surfaces—effectively a biological layer or living stratum. As a result, descendants of ancient Mesozoic life forms can exist more or less *in situ* and move onto much younger Tertiary strata. Life evolves biogeographically as if it were another geological stratum (fig. 2-5).

Case Study 1: Vicariant Disjunction along Fault Zones

One of the most prominent geological features of New Zealand is the Alpine fault zone. This feature exhibits extensive geological displacement and deformation over 480 km through Tertiary time (fig. 2-6a). Many taxa, including flowering plants, invertebrates, and vertebrates, exhibit disjunct affinities across this fault zone (fig. 2-6b). This disjunction is also evident in marine starfish and marine mollusca (fig. 2-6c). Movement along the fault can be correlated with these series of biological distribution patterns, which may have formed when whole communities of animals and plants were pulled apart by movement on the fault (Henderson 1985, Craw 1989, Heads 1990, 1994a, 1998).

California is well known as a parallel geological counterpart to New Zealand, sharing extensive fault zone displacement and associated earthquake activity. It has even been suggested that formation of the New Zealand Alpine fault zone was initiated by a late Oligocene Pacific-wide tectonic event as a consequence of the subduction collision of the East Pacific Rise with the North American continent (Kamp 1991). This event was significant in the Neogene development of California and triggered inception of the San Andreas transform fault system as a consequence of interaction between the Pacific and North American tectonic plates. Baja California and much of southwestern California have been displaced north–northwest for a considerable distance, from an original paleoposition in the Oligocene (30 MYBP) adjacent to western Sonora (Mexico). The shift from west to east of Baja California of interactions between Pacific and North American plates in the Pliocene (5 MYBP) resulted in the development of a spreading center and rapid separation of the Baja Californian peninsula from western Sonora, opening up the Gulf of California (fig. 2-7).

Formerly contiguous ancestral populations of taxa as ecologically diverse as coastal pines, freshwater fish, and reptiles may have been fragmented and pulled apart by these tectonic displacements. A freshwater cyprinid fish, *Gila orcutti*, of the Los Angeles Basin exhibits a major vicariant disjunction from its closest relatives, *G. ditaenia* and *G. purpurea*, of northwestern Sonora (fig. 2-8a). Mapping the present-day distributions on a palaeogeographic reconstruction of southwestern North

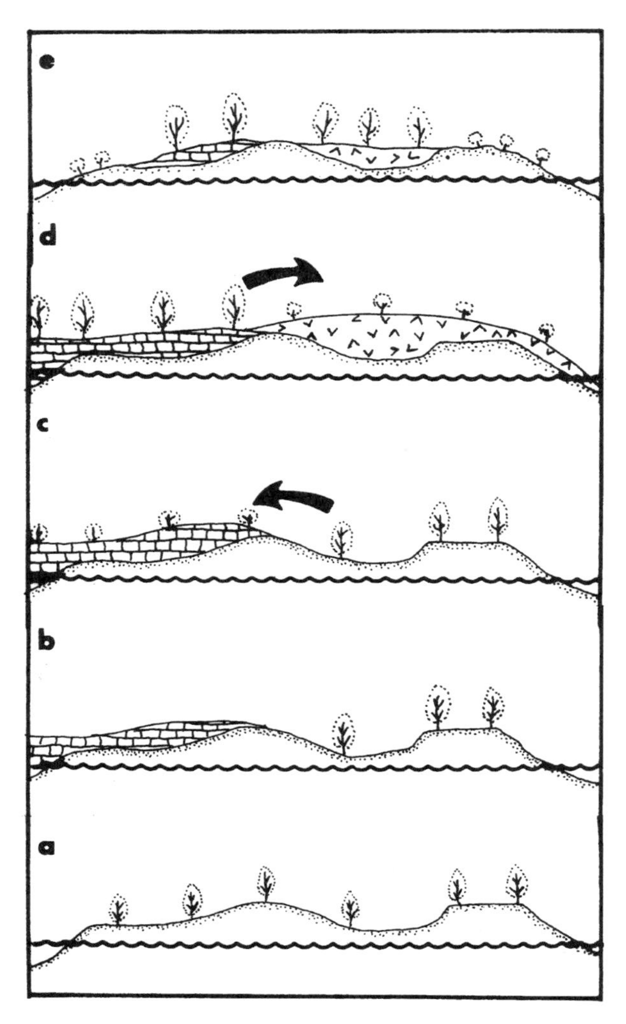

FIGURE 2-5. A model to illustrate the process of old life moving onto a younger geological surface. Through this process an old biota may remain more or less *in situ* through ecological means of survival allowing the communities to move over the locally changing landscape. (A) Mesozoic ancestral biota represented by trees on a late Mesozoic surface. (B) Mid-Tertiary marine transgression and subsequent uplift result in thick, exposed limestone sediments (horizontal and vertical hatching) over part of the Mesozoic surface. (C) This new landscape is colonized (curved arrow) by communities from the old peneplain surface. (D) Remnant Mesozoic peneplain covered by later Tertiary volcanic deposits (scattered v symbol), and this new surface is colonized (curved arrow) by communities from the mid-Tertiary limestone surface. (E) Late Tertiary erosion of the limestone and volcanic deposits allowing recolonization of the re-exposed Mesozoic peneplain surface by descendent communities of the original ancestral Mesozoic biota (from Heads 1990; reprinted with permission of SIR Publishing).

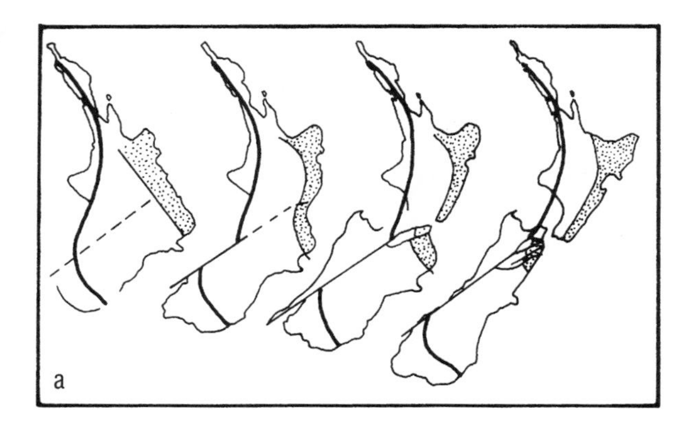

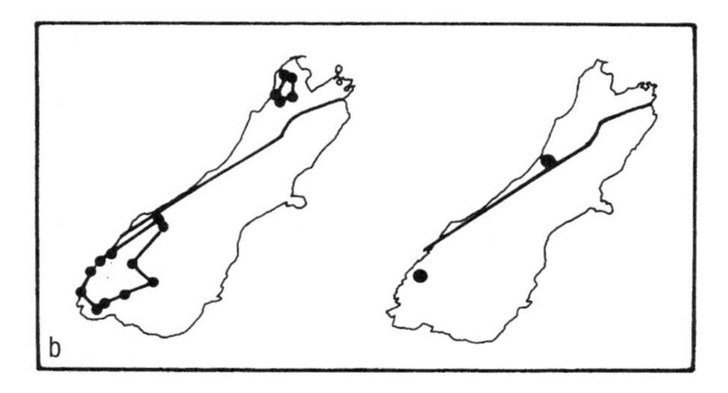

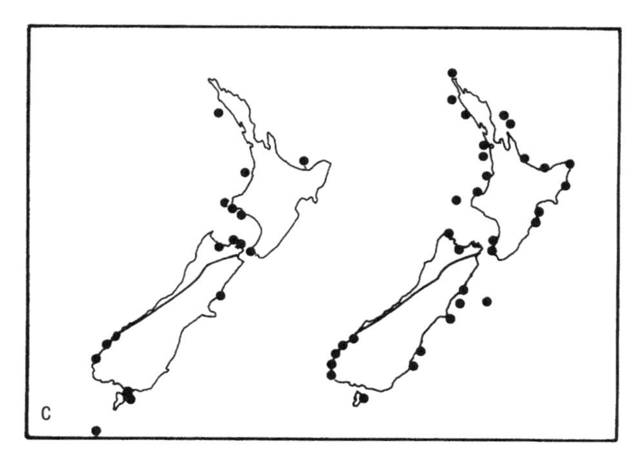

FIGURE 2-6. Correlations between Alpine transform fault and biological disjunctions in New Zealand (from Heads 1990; reprinted with permission of SIR Publishing). (a) Model of late Oligocene to present-day tectonic development of New Zealand; (b) disjunction of terrestrial taxa by Alpine fault zone (left, *Celmisia petriei* [Angiosperms: Compositae], right, the annelid worm *Deinodrilus benhami*); (c) disjunction of marine taxa by Alpine fault zone (left, *Evechinus chloroticus* [Echinoidea], right, *Divaricella* [Mollusca: Lucinidae]).

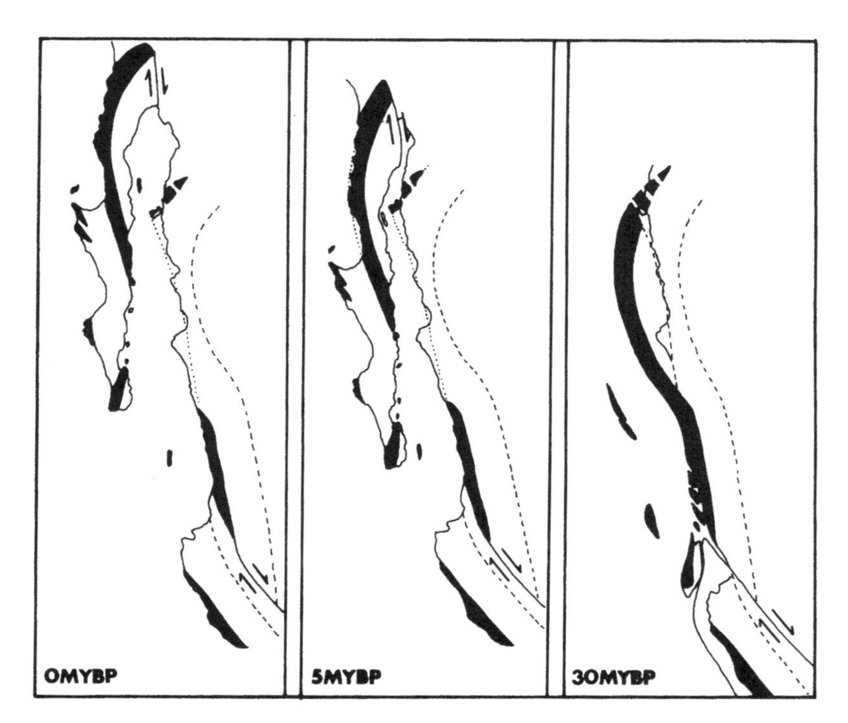

FIGURE 2-7. Model of Oligocene to present-day tectonic development of Baja California and southwestern North America due to movement on the San Andreas fault system. (Black shading) Granitic sub-belt; (arrows) fault movement (from Gastil and Jenskey 1973; reprinted by permission of Leland Stanford Junior University).

America prior to displacement of microplates along the San Andreas transform fault zone places these distributions in close juxtaposition (fig. 2-8a). Geological evidence supports contiguity of drainage patterns prior to tectonic displacement (Minckley et al. 1986). Similarly, Axelrod (1986) related disjunct distributions of closely related coastal pines of California and Mexico to northward translocation along the San Andreas rift system (fig. 2-8b). Murphy (1975, 1983a,b) suggested that the distribution of Baja California reptiles and isolation of endemic taxa on small islands in the Gulf of Mexico was correlated with fragmentation of their ancestral populations by formation of the gulf through these tectonic movements.

Case Study 2: Historical Legacy of Sea Level Changes

Rapidly changing configurations of land and sea may leave former coastal organisms stranded inland, and with tectonic uplift, coastal or

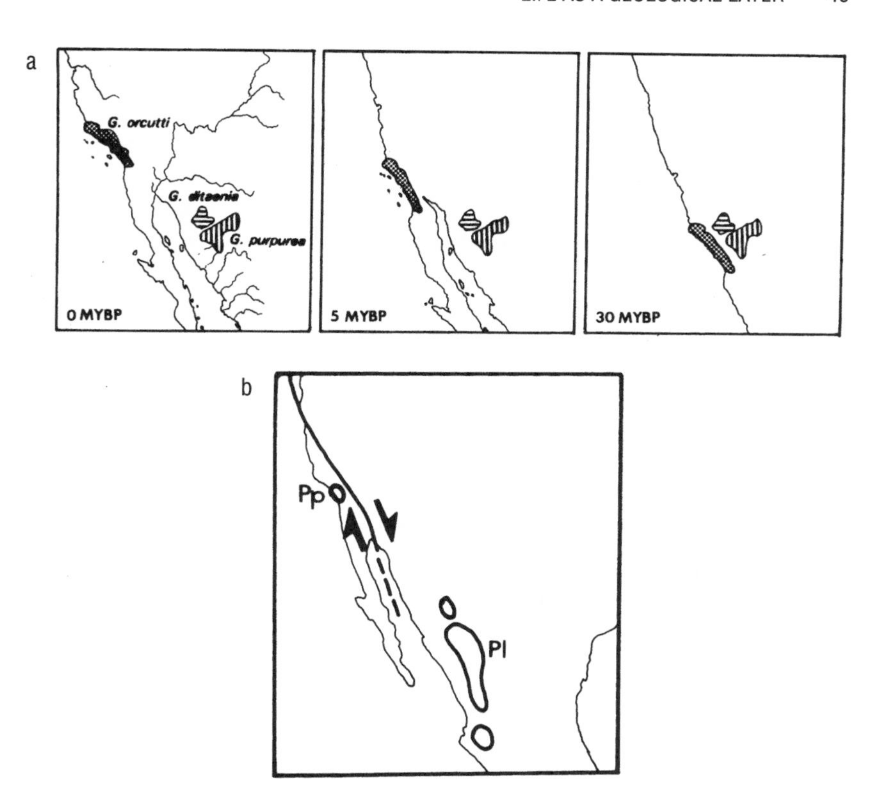

FIGURE 2-8. Correlation of disjunct distribution of closely related terrestrial taxa with movement of San Andreas fault system. (a) Left, present-day disjunct distribution of three closely related species of freshwater cyprinid fishes (*Gila orcutti*, Los Angeles Basin; *G. ditaenia* and *G. purpurea*, Sonora); center and right, these distributions mapped onto a paleogeographic reconstruction of southwestern North America, prior to displacement along the San Andreas fault system (from Minckley et al. 1986: fig. 15.9; reprinted with permission of John Wiley and Sons). (b) Disjunction of the closely related coastal pines *Pinus paucisquamosa* (Pp) and *P. lumholtzii* (Pl) on either side of the San Andreas fault system (after Axelrod 1986: fig. 15).

intertidal biota may eventually become inland, upland, or Alpine. Modernization of contemporary biota took place in late Cretaceous to mid-Tertiary paleoenvironments that were often lagoonal or estuarine and brackish to littoral. The coastal or intertidal marine origin of these organisms and communities can be described as "mangrove." This use of mangrove emphasizes the significance of the transitional zone between terrestrial and marine environments in the origin and diversification of many terrestrial organisms, and even the evolution of entire groups such as angiosperms (Croizat 1961, 1964).

Shallow-water coastal environments can be affected by comparatively slight movements of the geological shelf as well as eustatic sea-level changes. For example, shallow Tertiary (especially Oligocene) seas were spread widely over much of New Zealand, including localities that are now topographically montane. Survival of some biogeographical centers in New Zealand associated with earlier Tertiary events, despite sea level changes, suggests a complex, ever-changing system of inland shallow seas and islands. The occurrence of Oligocene beds in what are now mountains at elevations as high as 1600 m gives evidence of shallow embayments ancestral to the present Alpine landscape (Heads 1990).

Mesozoic basement rocks of the Central Otago region were deeply eroded and weathered in Cretaceous and early Tertiary time to a peneplain upon which was deposited a Cenozoic sequence. Much of this sequence was deposited as a consequence of marine transgression into the region from the east, south, and west, with maximum advance in the Oligocene. This was followed by Neogene uplift of the old peneplain and marine regression (Mortimer 1993). Some present-day inland communities of predominantly coastal plants and animals in Central Otago may be the result of stranding of former coastline biota by sea level regression, with subsequent survival of the coastal organisms being maintained by leaching of salt from the geological substrate to form localized salt pans (Patrick 1990).

One striking result of this marine transgression and regression is the formation of biogeographical tracks that encircle each other as a series of concentric rings. Distributional tracks of three Otago species in the plant genus *Leonohebe* (Scrophulariaceae) exhibit this type of pattern. The track of *L. propinqua* surrounds those of *L. poppelwellii* and *L. subulata*, in a concentric ring distribution (fig. 2-9a). These tracks may be correlated with both areas of sea-level regression such as receding inland seas and regions of progressive tectonic uplift. Oligocene Otago coastlines were probably sites of speciation, as weedy taxa invaded and evolved on the land exposed by shifting and gradually shrinking seas (Heads 1990, 1994c). Ecological and geographical vicariance of taxa along the shore line occurs sequentially as the water progressively diminishes, retreats, and finally disappears, leaving only the tracks to mark the former shoreline (fig. 2-9b).

Perhaps the most widespread recognition of correlation of distributions with sea-level changes occurs in biogeographical and ecological studies of groundwater organisms, where it is known as the "vicariance regression hypothesis." Many groundwater organisms are distributed in areas that were flooded by later Mesozoic or Tertiary seas, and the nearest relatives of these organisms are often found in marine situations rather than in surface freshwater environments. This relationship is explained as the result of marine shallow-water Crustacea being stranded

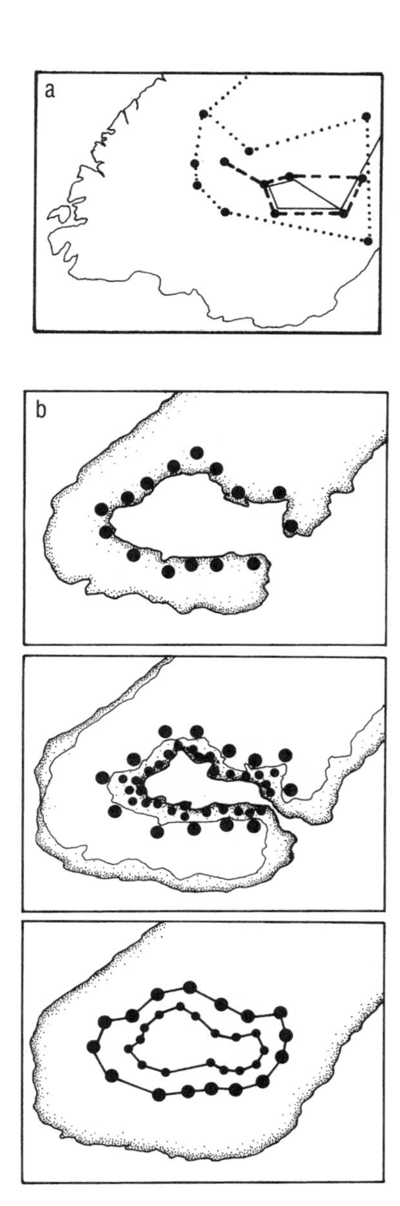

FIGURE 2-9. Vicariant regression and concentric ring distribution (from Heads 1990: fig. 23c; reprinted with permission of SIR Publishing). (a) Concentric ring distribution in southern South Island, New Zealand, for *Leonohebe* species. (b) Diagrammatic model for sequential form-making correlated with changes in topography of an epicontinental or inland sea. Top: ancestral distribution (solid circles) along shoreline; middle: weedy invasion of new habitat followed by differentiation through ecological vicariance (small solid circles) resulting in concentric ring distributions as organisms invade and evolve on land exposed by a shifting and gradually shrinking seas. Bottom, concentric tracks marking the former shorelines of now vanished sea.

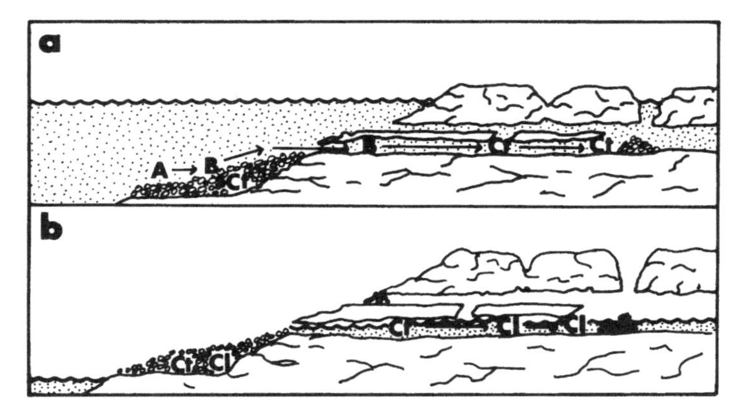

FIGURE 2-10. Vicariance regression model for marine-derived subterranean freshwater Crustacea. (a) During a period of marine inundation, benthic forms (A) may give rise to organisms that occupy crevices (B), which in turn may become marine crevice dwellers (Ct). (b) Through marine regression, these organisms may become trapped in newly developing freshwater habitats and evolve into underground freshwater taxa (Cl) descended from marine ancestors (after Holsinger 1994: fig. 7; reprinted with permission of Kluwer Academic Publishers).

during tectonic uplift or periods of low eustatic sea-levels (fig. 2-10). With a gradual lowering of salinity, the stranded crustaceans finally evolve into a groundwater fauna (Stock 1977, 1993; Holsinger 1986, 1991, 1994; Notenboom 1991). Marine regressions may be considered as vicariant events, but differentiation may be largely "polychotomous" rather than a simple dichotomous pattern of descent (Notenboom 1991).

Vicariance regression models are also applicable to the formation of groundwater fauna on volcanic islands that were formed from the seabed without any necessary direct contact with continental shorelines. Groundwater organisms on volcanic islands in the Caribbean, Galapagos, and Canary islands may include organisms that have been derived locally from widespread benthic marine ancestors. Historical events such as fragmentation or regression as documented by biogeographical studies could provide a biological calendar for geological events and provide a mechanism for dating an island's subaerial existence independently of paleontological or isotopic methods (Stock 1991).

2.4 Insular Distribution

The distinction of "oceanic" or "island" areas from "mainland" or "continental" regions is a driving assumption in many biogeographical studies, and is justified by the following arguments: (1) Islands provide clear examples of isolation, for the sea surrounding them is an environment in which few terrestrial or freshwater organisms can survive. Special adaptations for transport by air or water are necessary for an organism to cross a stretch of ocean, so dispersal to islands is by a "sweepstakes" route, with the successful organisms sharing adaptations for crossing the intervening region rather than for living within it (Cox and Moore 1993). (2) The oceanic island has a *"de novo"* origin without any direct connection to a mainland. Evidence for this isolation is traditionally derived from geological theory. Petrology of the basalts that form these islands indicates that they are marine in origin. The Galapagos and Hawaiian archipelagoes, for example, are classic examples of oceanic islands. With their geographic isolation and obvious geological origin as submarine volcanoes, it would be unreasonable to think of them as anything else (Keast 1991).

Many islands show a biogeographical pattern where islands share closer biogeographical relationships with each other than with the nearest mainland, even though the interisland distances may greatly exceed the distances between individual islands and the mainland. The islands behave biogeographically as if they were part of a collective unit that belies their apparent insularity. When these biogeographical relationships are mapped as a track, the result is a biogeographical arc that parallels the mainland (fig. 2-11a), or sometimes includes the mainland at the track margins (fig. 2-11b). In these patterns the islands are "in front" as the biogeographical focus, while the mainland is relegated to the background as if it were secondary or irrelevant to the distribution. Such distribution patterns are contrary to what might be expected if the island biota were derived by long-distance jump dispersal over water from the closest continental source areas. Contrasts between islands and mainland may be attributed to contrasting ecology, but the interisland patterns are often complex, involving organisms with different means of dispersal, different ecological requirements, and often without any apparent ecological contrasts with nearby mainland habitats. The interisland tracks may also connect with certain parts of the mainland and not others, despite the presence of suitable habitat throughout the mainland.

With plate tectonic movement, original geologic formations may be removed and accreted or subducted elsewhere, so that former insular distributions can become continental through island arc accretion to mainland areas. Tyler (1979) correlated frog distributions in New Guinea with island arc accretion, and Van Welzen et al. (1992) related collision and docking events involving numerous island arc terranes in northern New Guinea to the diffusion and vicariant differentiation of ancestors to the modern biota. Citing numerous examples based on the Tuscan Pliocene archipelago in Italy, Lanza (1984) discussed the impact of fossil islands on the distribution of extinct endemic taxa. Fossil is-

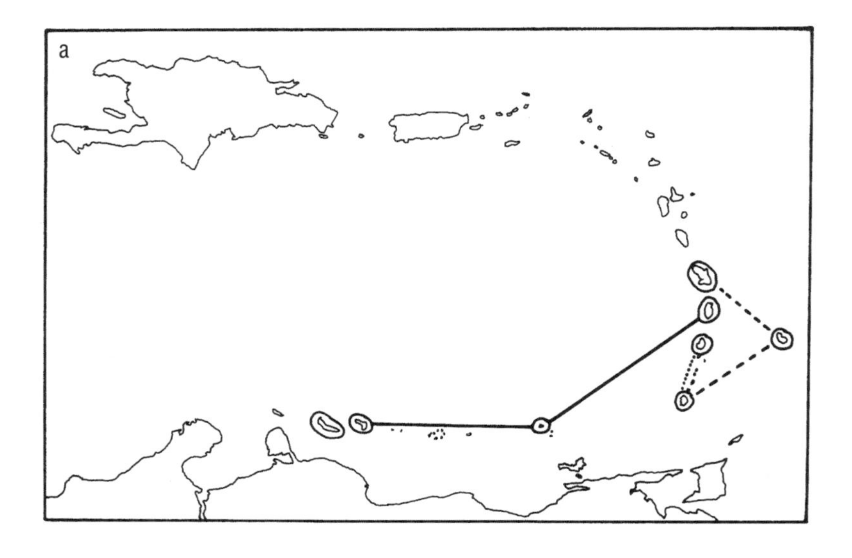

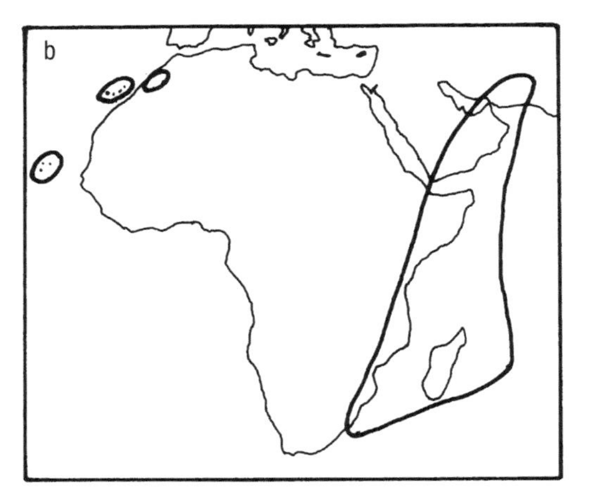

FIGURE 2-11. Insular distribution in relation to neighboring mainland. (a) The lizards of the *Anolis* species group on the Lesser and Netherlands Antilles, Caribbean Sea (data from Roughgarden 1995); (b) the plant tribe *Sideroxyleae* (Sapotaceae) between the Cape Verde and Canary Islands and the North African mainland at Morocco (after Aubréville 1974b: fig. 88).

lands are those that have ceased to exist due to their fusion, through marine regression and tectonic activity, with a continental area or with one or more islands. Although the islands no longer exist as discrete entities, they are still represented by the local endemic fossil distributions of their former fauna and flora within the new landscapes.

The Pacific Ocean is littered with oceanic islands and island chains,

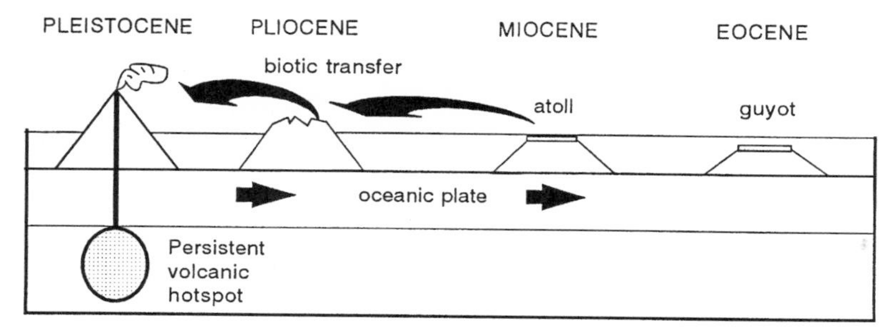

FIGURE 2-12. Survival of an older insular biota of Eocene age on younger Pleistocene and Pliocene islands in an island chain as new islands form at a persistent hot spot and provide new habitats that are colonized from adjacent older islands. As the islands move away from the hot spot, they are subjected to erosion until they finally submerge beneath the sea or ocean, but their original biota or its descendants continue to survive *in situ* near the hot spot on younger emergent islands (after Axelrod 1972: fig. 25; reprinted with permission of University of Arkansas Museum).

but many are associated with long-term volcanic hot spots that continue to form new islands as the old ones are rafted away and eroded down to form coral atolls, and eventually submarine seamounts. For organisms to maintain viable populations over an ever-changing archipelagic topography of emerging and submerging islands, they require effective means of survival that enable populations to reestablish on new habitats as the older ones erode and disappear (fig. 2-12). Many island organisms possess an easily dispersed phase in their life cycle ecology necessary for survival in these constantly changing insular chains and did not necessarily migrate over water to their present habitats (Axelrod 1972).

Case Study 1: Ancient Life on Young Islands

The Hawaiian Islands are part of a chain of islands and seamounts that extend back at least 70 MYBP, and high altitude islands have existed in the vicinity of the Hawaiian Islands since late Paleocene time. There is considerable evidence of biotic movement from older to younger islands, and some evidence exists of ancient terrestrial biota, with the possibility that some of the islands' biota is derived from even earlier islands (Lewin 1985, Carson and Clague 1995, Wagner and Funk 1995). These islands have a native land-snail fauna of more than 750 described species, most of which are endemic to the archipelago. This species richness appears to be related almost entirely to *in situ* vicariant evolutionary differentiation, rather than to an equilibrium between immigration and extinction. Ancient in origin, the Hawaiian land molluscan fauna has evolved on both present-day and ancestral submerged or subducted islands (Cowie 1995). The estimated 1,000 species of endemic Hawaiian drosophilid flies (Drosophilidae) have been characterized as extraordinary and remarkable (Kaneshiro et al. 1995). Varied ages ranging from 40 to 23 MYBP have been calculated for the divergence time of these endemic flies from their North American relatives. These estimates all suggest that the group is older than the oldest present-day high island in the archipelago (Powell and DeSalle 1995).

The now low-island Hawaiian atolls of Nihoa, Necker, and Laysan, ranging in age from 7.2 to 15 MYBP, were once large volcanoes high enough to sustain rainforest, which have now eroded almost down to sea level. Laysan and Nihoa retain (or did until recently) numerous genera of upland and montane lineages of animals and plants characteristic of high volcanic islands in the chain rather than atolls. These include the palm *Pritchardia*, two genera of Hawaiian honey creepers (Aves: Drepanididae), an endemic duck (*Anas platyrhynchos laysanensis*), and the land snail *Tornatellides*. They are most plausibly regarded as a relict biota, millions of years old, rather than as recent arrivals through long-distance dispersal from younger high volcanic islands elsewhere in the Hawaiian chain (Schlanger and Gillett 1976, Carson and Clague 1995).

Some islands and seamounts in the Hawaiian chain are significantly older than others (e.g., Wentworth Seamount at 70 MYBP) and are postulated to have moved with the Pacific plate from a position south of the equator. A mixing of Hawaiian and non-Hawaiian terrestrial biota could have resulted, if they were above water when first arriving near the Hawaiian hot spot (Rotondo et al. 1981). This southern relationship is supported by the distribution of a number of organisms (see Springer 1982). Thus, components of the Hawaiian biota may not only be much older than the current islands, but also have first evolved within a completely different palaeogeographical context.

Case Study 2: Evolution of the Galapagos Biota

Isolated from the South American mainland by almost 1000 km of open sea, the Galapagos Islands have interested evolutionists ever since they first came to the attention of Charles Darwin. Darwin (1859) suggested that the biota was derived by long-distance dispersal from the American mainland. This has remained a popular scenario because the current islands are young (from 1 to about 4 million years old) and have arisen from the sea floor through volcanic activity. The Galapagos finches, for example, are characterized as originating from an "epic" dispersal colonization event exemplifying the role of stochastic factors in distribution (Haydon et al. 1994b).

Many Galapagos species are widespread elsewhere in the eastern Pacific and on the American mainland. Others are endemic to the islands, and many of these are related to taxa on the American mainland. The geography of these relationships includes three vicariant track patterns that meet at the Galapagos Islands node. This archipelago occupies a position at the intersection of tracks running north to Baja California and other parts of North America, north–northeast to Central America and the Caribbean, and east to western South America. Track relationships with South America often involve taxa with distributions occurring along or near the Pacific coast of the continent. For example, numerous endemic species of the flightless weevil genus *Galapaganus* occur in the Galapagos Islands, and several related species occur in Ecuador and Peru (fig. 2-13a).

Central American and Caribbean tracks can be illustrated by *Paracance* (Diptera: Canaceidae) flies with Galapagos–Panama–Brazil, and Cocos Island–Dominican Republic tracks (fig. 2-13b). The endemic Galapagos species in this genus are the least derived taxa for the two lineages (Mathis and Wirth 1978). An endemic geometrid *Oxydia* moth on Cocos Island is most closely related to a Lesser Antilles species, and the genus is also represented by an endemic Galapagos species (fig. 2-13c). This distribution pattern is considered to be consistent with the possibility that the islands were once part of a formerly contiguous eastern Pacific–Caribbean are fragmented by the early Cretaceous break-up of a proto-Antilles archipelago (Brown et al. 1991). The Caribbean affinities of some Galapagos and Cocos island tineid moths indicate that they are components of this ancestral biota (Davis 1994). Numerous marine invertebrate and terrestrial vertebrate taxa are also components of this Galapagos–Central America–Caribbean track (Rosen 1976).

A possible northern track relationship is illustrated by the endemic cotton plant *Gossypium klotzschianum*, which may be most closely related to a species in Baja California (fig. 2-13d). In contrast, the Peruvian cotton, *G. raimondii*, is either more closely related to *G. gossypioides* (Wendel and Albert 1992), indicating a parallel track, or it is the sister group of a Hawaiian and New World species group, which includes the Galapagos endemic (Wendel et al. 1995). Despite uncertain and continually changing cladistic tree topologies for cotton relationships, the Galapagos cotton is nodal for two vicariant track patterns. A zoological example of this northern Galapagos track is provided by the endemic iguana genera *Amblyrhynchus* and *Conolophus*, which appear, on the basis of molecular and morphological data, to be most closely related to *Ctenosaura* from Mexico and Honduras (fig. 2-13e) (Sites et al. 1986). An alternative hypothesis of relationship for the Galapagos iguanas places them in a monophyletic clade with the chuckwallas (*Sauromalus*) of the desert regions of northwestern Mexico and southwestern United States (de Queiroz 1987). Although taxonomically significant, this different

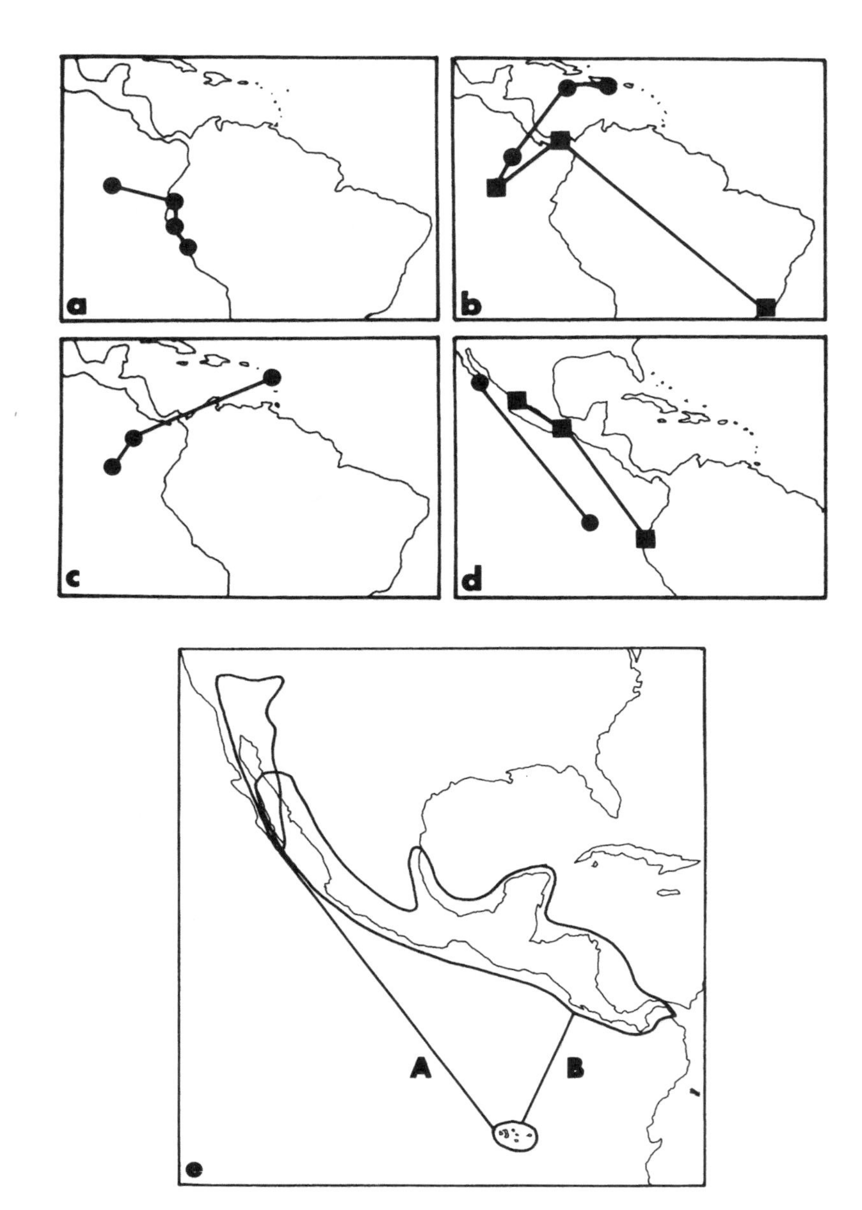

FIGURE 2-13. Examples of standard tracks for taxa found in the Galapagos Islands. Central American-Caribbean tracks: (a) *Galapaganus* (Coleoptera: Curculionidae) (data from Lanteri 1992); (b) *Paracance* (Diptera: Canaceidae) flies (data from Wirth 1956, Mathis and Wirth 1978); (c) *Oxydia* (Lepidoptera: Geometridae) (data from Brown et al. 1991); (d) Northern Tracks: *Gossypium* (Malvaceae) species (classification from Wendel and Albert 1992, Wendel et al. 1995); (e) Galapagos iguanas and their Mexican and Honduran relatives. Track A, classification of de Queiroz (1987); track B, classification of Sites et al. (1996).

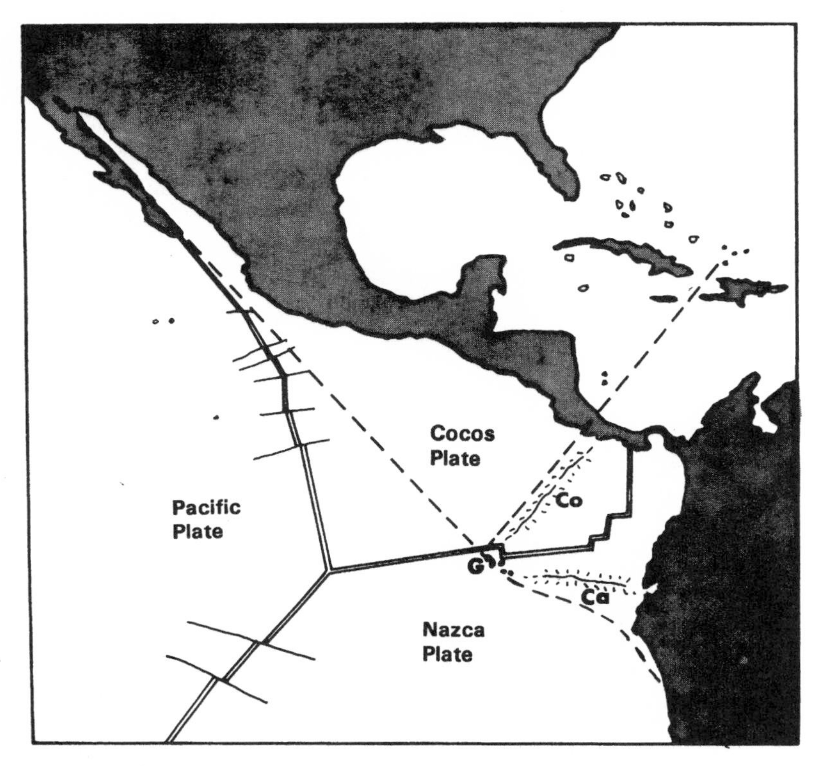

FIGURE 2-14. Correlation of tracks and tectonics for the Galapagos Islands. Note the relationship of the standard tracks (dashed lines) overlapping in the Galapagos Islands (G) to the tectonic junctions associated with the intersection of the Nazca, Cocos, and Pacific lithospheric plates, and the Cocos (Co) and Carnegie (Ca) ridges.

phylogenetic perspective does not alter the geometry and orientation of their track, in that either relationship supports a northern track.

These predominantly vicariant standard tracks that overlap in the Galapagos Islands represent a similar biogeographical incongruence regarding means of dispersal and geography to that occurring on continental areas. These three standard tracks may be correlated with the position of the Galapagos near a tectonic junction involving the intersection of the Nazca, Cocos, and Pacific lithospheric plates and the Panama fracture zone (fig. 2-14). These tracks are "continental" despite having a partially "oceanic" setting and imply a former terrestrial landscape nearby what is now an oceanic volcanic hot spot.

The Galapagos Islands have been characterized as the Achilles heel of panbiogeography because they are volcanic islands that arose *de novo* from the sea floor, and shallow marine and terrestrial Galapagos species possess adaptations for long-distance transport (Briggs 1991, Keast 1991). As Mayr (1982b:619) comments: "If panbiogeography could be so wrong about the Galapagos, how could it be right about anything else?" However, estimated sea floor spreading rates by Holden and Dietz (1972) imply progressively older sea floor to the east of the current Galapagos Islands, reaching an age of 40 million years at the subduction zone on the American coast, and they suggest that the modern Galapagos Islands may have inherited their biota from a series of ancestral Galapagos Islands over a 40-million-year period. Cobbles collected from underwater seamounts on the Cocos and Carnegie ridges have been interpreted as the products of erosion near sea level, indicating the past existence of volcanic islands now submerged beneath the ocean (Christie et al. 1992). Geological observations and radiometric data suggest islands were present at the Galapagos hot spot for at least 9 million years, and the tectonics of the Cocos and Carnegie ridges are consistent with hot-spot volcanism for 15–20 million years. Because large portions of the Caribbean plate are linked to the initiation of Galapagos volcanism, it is likely that islands have existed throughout the 80- to 90-million-year history of hot-spot activity (Pindell and Barrett 1991, Carson 1992, Glaubrecht 1992).

Historical inferences from track and tectonic correlations may not necessarily agree with current or prevailing paleogeographical reconstructions, but in the case of the Galapagos archipelago there is some correspondence. Plate tectonic models for the origin of the Caribbean region generally recognize an eastward movement of the Caribbean plate relative to North America, implying a 1000–1400 km offset since the Eocene. This would place the present Central American region well into the Pacific as late as the mid-Tertiary (Duncan and Hargraves 1984, Mattson 1984, Durnham 1985, Pindell and Barrett 1991). Some of these models postulate a direct connection between Mexican and Peruvian terranes before separation in the middle to late Jurassic (Pindell and Barrett 1991, Mann 1995), an eastern Pacific location for the Panama–Choco island arc until its collision with South America in late Miocene to Pliocene time (Kellog and Vega 1995), and a Pacific origin for the Caribbean plate (Stockhert et al. 1995). A possible paleogeographic relation between western North American cordilleran terranes and Jurassic fragments (including a remnant seamount and island arc) of Pacific origin now situated along the northeastern margin of the Caribbean plate has been suggested (Montgomery et al. 1994).

Many areas of the present-day Central and South American mainland, such as Panama, western Ecuador, and northwestern Peru, that are often suggested as centers of origin for Galapagos founder populations, appear to be a collage of island arcs, seamounts, and microcontinental fragments that accreted to North and South America through late Mesozoic and Tertiary time (Feininger 1987, Mourier et al. 1988, Laj et al. 1989, Di Marco 1994, Mann 1995). For instance, there are 10 endemic flightless *Galapaganus* weevil species on the Galapagos Islands, variously related to five *Galapaganus* species distributed along the northwestern South American coast from Guayas Province and the Guayaquil Gulf region of western Ecuador in the north to Lima and Callao on the central Peruvian coast in the south (fig. 2-13c). Lanteri (1992) considered that the cladistic relationships of these weevils indicated two or three colonizations of the islands from South America, but the distributions of three out of five of the mainland species lie largely within the boundaries of an accreted allochthonous island arc and associated sediments in western Ecuador, while the central and northern Peruvian coastal species occur mainly in a region that was a volcanic and plutonic belt in the Cenozoic. Distributional fidelity to terrane boundaries is exhibited by *G. propinquus*, *G. howdenae*, and *G. femoratus* in Ecuador, whose geographic ranges are circumscribed by the Piñón terrane (fig. 2-15). Lithologies of the Piñón formation are geochemically similar to mid-oceanic ridge basalts from the Galapagos Rise (Laj et al. 1989), and this terrane has been interpreted as having become fixed to cratonic South America in late Cretaceous to Eocene time coincident with the collision of the Macuchi Island arc (Feininger 1987, Kellog and Vega 1995). Farther south in northwest Peru, *G. squamosus* exhibits some geographic fidelity with the Amotape-Tahuín allochthonous continental terrane as its northernmost record (Chilcayo) lies within that microcontinental fragment's boundaries (fig. 2-15). Remaining records of this species (Campana and Trujillo) and those of *G. lacerotosus* from Lima and Callao on the central Peruvian coast are more problematic with respect to terrane fidelity, though Feininger (1987) notes that lower Cretaceous rocks on the coast at Lima may have had a westerly continental source in an area now occupied by the Pacific Ocean.

Biogeographical models treating the Galapagos group as isolated oceanic islands populated recently by actively or passively dispersing organisms are not necessary given such paleogeographic and tectonic constraints. Extant Galapagos animals and plants, whether weedy or relictual, could have originated on a series of forelands and island arcs formerly occupying the eastern Pacific, with descendant biota surviving at this tectonic hot spot, possibly throughout the last 80 million years (fig. 2-16). Inheritance of biota older than the present islands is documented for the land and marine iguanas, with immunological studies estimating a divergence time of 15–20 MYBP compared with an age of 4 MYBP for the oldest current island (Wiles and Sarich 1983). This estimate, together with the biogeographical tracks and tectonic models, provides a threefold corroboration pointing to an origin for at least part of the Galapagos biota that pre-dates the origin of the present islands, not on modern mainland America, but in an area that is now the eastern Pacific Ocean (Croizat 1976, Roughgarden 1995).

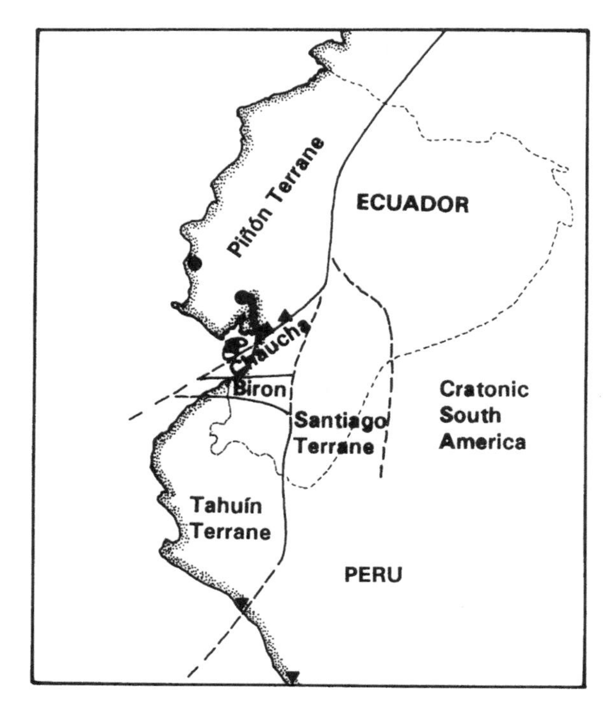

FIGURE 2-15. Distributional fidelity of South American *Galapaganus* weevils with boundaries of allochthonous terranes (Piñón, Biron, Chaucha, Tahuín) in western Ecuador and northwestern Peru. (Circles) *G. howdene*; (square) *G. propinquus*; (triangle) *G. femoratus*; (inverted triangle) *G. squamosus*.

2.5 Conclusions

Correlation of biological and geological patterns suggests that the evolution of individual taxa and communities is strongly mediated by vicariant and dispersed events associated with earth history. The evolution of individual species, ecological communities, and even entire biota, can be represented as a series of unified processes without any necessary discrimination between the geological and biological components. Correlations involving organisms as diverse as freshwater fishes, pine trees, and reptiles can be recognized as individual cases of a general relationship involving earth and life evolving together, and panbiogeography offers general principles and methods for integrating these events within a broader understanding of ecological communities in general.

Modern ecological communities are evolutionary composites at the intersection of regional and global tracks correlated with earth history extending back to late Mesozoic and early Tertiary times. Contiguous

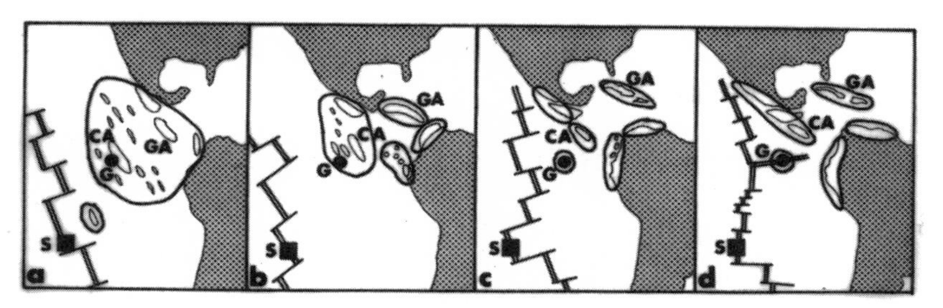

FIGURE 2-16. Generalized historical model to illustrate the potential biogeographic relationship between biotic distributions (solid lines) associated with former eastern Pacific terranes/island arcs in relation to the Galapagos hot spot (filled circles) and American mainland (shaded). (a) 85 MYBP, (b) 60 MYBP, (c) 38 MYBP, (d) 7 MYBP. CA, Central America; G, Galapagos hot spot; GA, Greater Antilles; S, Sala y Gomez hot spot. This model represents a compilation of several geological reconstructions (see text for details) to show how the present-day Galapagos Islands could inherit a sample of eastern Pacific organisms when a series of island arcs moved over or within the vicinity of the Galapagos hot spot during the Mesozoic and Tertiary. The biogeographic and geological relationship should provide a focus for future geological and biogeographic research into the origins and evolution of the Galapagos biota.

ecological communities on current landmasses comprise tectonic and biological composites formed through the historical suturing of different biogeographical histories, and this suturing may have profound geological, geographical, and biological effects, including the evolution of biota along Cretaceous and Tertiary shores and the formation of high elevation communities from lowland coastal communities through tectonic uplift. The evolution of biological diversity is integral with earth history, and tectonics provides an evolutionary context for ecological and phylogenetic differentiation and diversification. Rocks, landscapes, and biota are metamorphosed as one.

3

Ecology, History, and the Panbiogeography of Africa

Approaches to biogeography have been and remain many and varied, but in essence either an ecologically or historically focused approach is adopted. Conventionally, if the temporal, spatial, and systematic scales are broad, then the focus is historical biogeography, if narrow then the subject is ecological biogeography (Myers and Giller 1990). Subdivision into plant (phytogeography) and animal (zoogeography) biogeography is also often made, especially in the older literature.

The term "ecological plant geography" can be traced back to the Danish botanist Eugenius Warming's classic text *Plantesamfund Gruntrak af der okologiska Plantgeografia* (1895) and was consolidated in the influential work of Schimper (1898) (see Fosberg 1976 and Stott 1984 for excellent reviews of this field). Ecological animal geography originated in the work of the German zoologist Richard Hesse (1913, 1924) and was characterized, in contrast to historical zoogeography by Hesse et al. (1937:6–7):

> Historical zoogeography . . . attempts to work out the development in geologic time of present-day distribution by studying the homologies of animal distribution. For such studies the starting points may be the systematic groups of animals. The ecological viewpoint, as contrasted with the historical, regards the analogies between animal communities in similar habitats. . . . Ecological zoogeography views animals in their dependence on the conditions of their native regions, in their adaptation to their surroundings, without reference to the geographic location of this region.

Although this division seemed established by the 1960s, ecological biogeographical analyses have since begun to emphasize the historic, the homologous and the systematic (Cadle and Greene 1993, Brown 1995). Using the birds of the Mediterranean area, and in particular the islands of Corsica as case studies, Blondel (1988) showed, for example, that "many features of community ecology can only be properly understood if interpreted in the light of history." Similarly, Vuilleumier and Simberloff (1980) have noted that "both ecology . . . and history . . . have played roles together at all times: they are indissolubly tied up, and there is probably no way to separate them, for nature is not dichotomous as far as 'time' goes. Ecological time and evolutionary time actually merge." These and other comments by prominent community ecologists and ecological biogeographers (e.g., Wiens 1989, Ricklefs and Schulter 1994) demonstrate that the customary and long established divisions between ecology and history in biogeography can no longer stand.

That space and time are central to the construction of all ecological interaction does not mean that ecological theory must be historically and geographically specific as called for in the program of historical ecology (Brooks 1985, Brooks and McLennan 1991). Rather, ecological theory should be about the geographical and historical constitution of ecological structure right from the start. Yet to date most empirical studies still treat regions and areas of endemism as simply unproblematic and untransmutable containers for ecological processes. This idea of regions and areas as unproblematic means that any distribution patterns undermining the presumed integrity of the biogeographical units have to be ignored, classified as special case anomalies involving taxonomic error (i.e., are the result of convergence or parallelism), or attributed to unique events unrelated to general biogeographical and ecological patterns. These issues are examined using examples from African biogeography analyzed according to the method outlined in chapter 1.

3.1 Regional Biogeography of Africa

For nineteenth-century European fiction writers, Africa represented something like the unconscious. But for Western biology, Africa and the tropics have always provided direct stimulus. It was a Guyanese African, John Edmonstone, who first enthused Darwin about tropical rainforest, and this was Darwin's inspiration for travel and ultimately for his theory of evolution (Desmond and Moore 1992). Biogeographers of the time saw Africa as the "Ethiopian region" that was subdivided into three regions: a central region of high, closed rainforest (Zaire, etc.), a southern center of diverse shrublands around the well-known Cape of Good Hope, and the remainder of the continent—a vast area of woodland, savannah, and desert.

These regions are seen in Wallace's (1876) writings (fig. 3-1), and are slightly modified in more recent texts (e.g., Brenan 1978). Unfortunately,

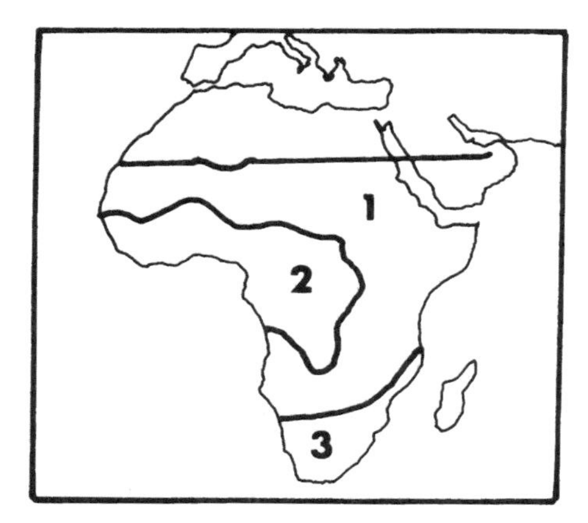

FIGURE 3-1. The nineteenth-century concept of the African continent (south of the Sahara) represented by Wallace (1876) as a composite of three regions: 1, a vast area of woodland, savannah and desert; 2, the central rainforest region, and 3, the southern region of diverse shrublands. These boundaries obscure important differences within forest and savanna regions, and important structural and phylogenetic affinities between these regions and other lands.

they prevent a deeper understanding of geological and biological history by obscuring important differences within the forest and savanna regions, as well as important structural and phylogenetic affinities between those regions and other lands. The flora and fauna of East, West, and South Africa are intimately related with those of Asia, Brazil, and Australasia, respectively, and proposed "elements" in the African biota, such as the "Afromontane" flora, are neither especially African nor exclusively montane. Furthermore, widespread "African" taxa are often notably absent from areas such as North Africa or southern Africa and Madagascar.

Early in the history of biogeography, the southern tip of Africa around the Cape was recognized as a major center of biodiversity, along with Australasia and Patagonia. These three southern centers were often explained as the result of immigrant species from the Northern Hemisphere piling up in the dead-ends of South Africa, South America, and Australasia (Bews 1921). This theory of biotic dominance was applied to the Cape genera by Levyns (1964:94), who believed that present-day distributions show the repeated "picture of a genus arising in the north, leaving behind scattered records of its journey southward, and finally undergoing intensive speciation in the southwest." Similarly, it has been suggested that the "montane rainforest" from South Africa through East Africa to Ethiopia originated in northern Africa (Algeria) and migrated south (Axelrod and Raven 1978). Likewise, Lawes (1990) argued for a radiation of the samango monkey "into southern Africa" from the north, and Davis (1993) envisioned the "penetration" of certain groups of scarab beetles into the southwest Cape from the north.

Despite widespread acceptance of this tradition of African biogeography, an alternative view has existed ever since biogeography came into existence as a global science. Already in the eighteenth century, there were suggestions of primary biogeographical affinities between the Cape region of South Africa and Australia (Wildenouw 1798, cited in Weimarck 1934). Hutchinson (1946) interpreted the southern African–Australasian connections as supporting Wegener's hypothesis of drifting continents, and criticized the idea of the European origin of southern African plants. Boughey (1957) noted the extensive evidence for a Southern Hemisphere focus for much of African life. Wild (1964, 1968) argued that present-day distributions in Africa are fragmented former Gondwana distributions and that explanations based on long-distance migrations were unnecessary.

3.2. Tracks and Baselines of African Biota

African plants and animals illustrate two major patterns of transoceanic tracks and baselines: Indian and Atlantic, and a third (Indian/Atlantic) which combines them. Numerous examples of these relationships are illustrated in the literature, especially for plant (e.g., Croizat 1958, 1968b; Aubréville 1969, 1974a,b; Axelrod 1972) communities that demonstrate the intercontinental structure of African ecology. The region Madagascar—the Cape–East Africa, in particular, marks one of the preeminent nodes of modern plant and animal life—the African gate of angiospermy. Biogeography demands the attention of the evolutionary ecologist because the origins and composition of rainforests and other "internal" African habitats are integral with the biogeography of ecological communities beyond Africa. In the following section, a few examples are presented to illustrate the application of the track method and the baseline criterion in classifying the biogeographical components that make up ecological communities in the African biota.

Indian Ocean Baselines

Biogeographical connections of African organisms across the Indian Ocean have sometimes surprised animal and plant geographers. Jeffrey (1988) noted the "astonishing geographical disjunction" of *Dactyliandra welwitschii* (Cucurbitaceae) between the deserts of southwest Africa and Rajasthan (India). On finding a continental Asian moss on Madagascar, Touw (1993) was astonished. These disjunctions are remarkable only for their lack of geographic continuity, and they are otherwise common-

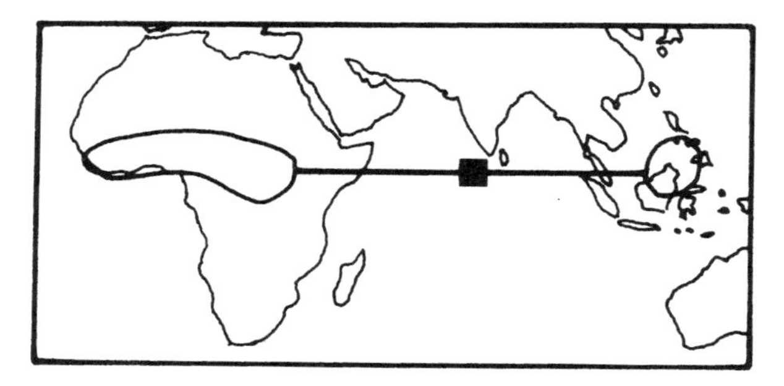

FIGURE 3-2. Indian Ocean baseline and generalized range of *Airyantha* (Leguminosae) illustrating a classic African–Indonesian disjunction (from Brummitt 1968). This pattern is classic in that it is common to many other organisms and highlights the close biogeographic interrelationships of West Africa and Southeast Asia.

place distributions. The legume *Sphenostylis*, for example, is widespread in tropical and southern Africa, but a single species also occurs between Bombay and Goa on the Indian coast (Gillett 1966). Similarly, the West–Central African *Airyantha* is also present in northern Borneo and southern Philippines (fig. 3-2).

Baobab trees (*Adansonia*) and geraniums (*Pelargonium*) represent two vicarious African relationships with Australia. The baobabs are widespread in Africa and Madagascar (with most species), but localized in northwestern Australia (fig. 3-3a). In contrast, *Pelargonium* occurs in southern and northern Africa, as well as on some southern islands and in southern Australasia (fig. 3-3b). More widespread at both ends of the Indian Ocean basin is the tribe Gnidieae (Thymelaeaceae) with *Gnidia* in Africa (especially southern Africa) and India, and three more genera in Borneo, the Philippines, Australasia, and the southwest Pacific (fig. 3-3c). Groups of birds with Indian Ocean baselines include shelducks (*Tadorna*) and austral teals (*Anas*) (Southey 1990), falcons (Falconidae) (fig. 3-4a), swifts (Apodidae) (fig. 3-4b), and Accipitridae (fig. 3-4c).

Alternating patterns of relationship may occur in different continents, such as the overlapping distributions of southern African *Caesia* and *Roridula* that are vicariant with *Papaver aculeatum* (fig. 3-5). In Australia *Caesia* is widespread and overlaps the vicariant and localized ranges of *Byblis* in the southwest and northeast (but vicarious to the Australian baobabs), and *P. aculeatum* in the southeast.

Other Indian Ocean distributions may involve only the eastern regions of Africa proper, or Madagascar and Indian Ocean islands. *Givotia* occurs on the mainland only in Somalia (fig. 3-6a), whereas *Athrixia* ranges from the Cape to the Red Sea, as well as in southwestern Australia (fig. 3-6b). Organisms as different as moths, bats, palms (*Chrysali-*

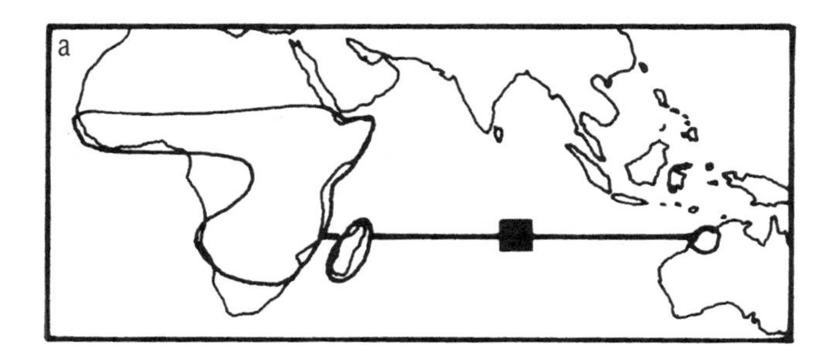

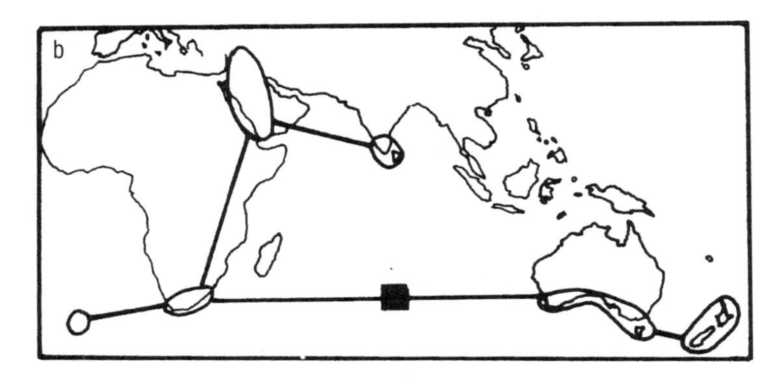

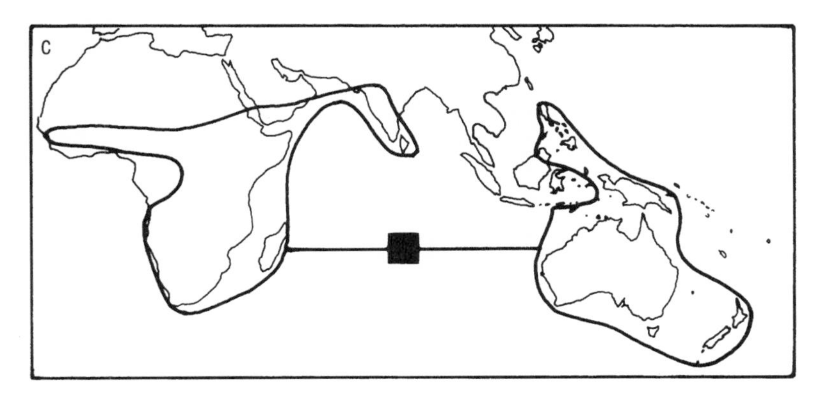

FIGURE 3-3. Indian Ocean baseline and generalized range of (a) *Adansonia* (Bombacaceae) (from Armstrong 1983, Baum 1995, Wickens 1983), (b) *Pelargonium* (Geraniaceae), and (c) Gnidieae (Thymelaeaceae) (from Heads 1990, 1994d). Note how the distributions of *Adansonia* and *Pelargonium* share a common biogeographic homology with the Indian Ocean while emphasizing different geographic sectors to the north and south of Africa and Australia. In contrast, the Gnidieae are more widespread at both ends of the Indian Ocean.

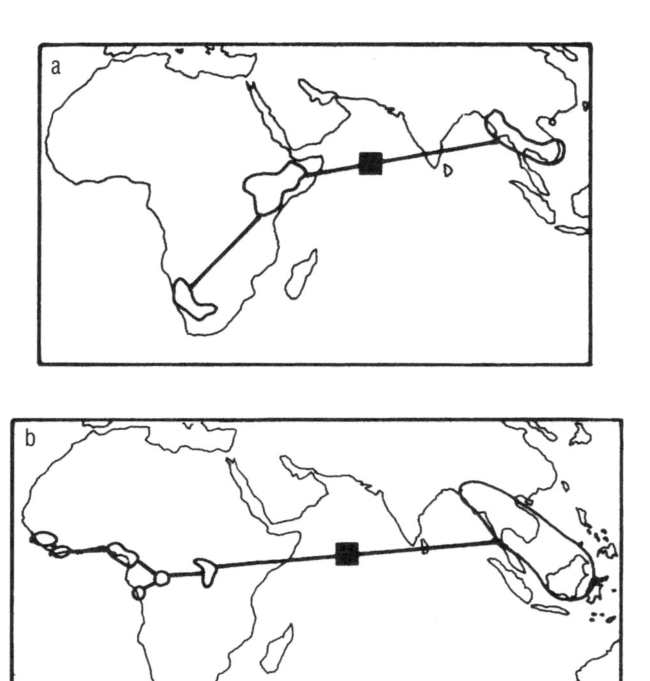

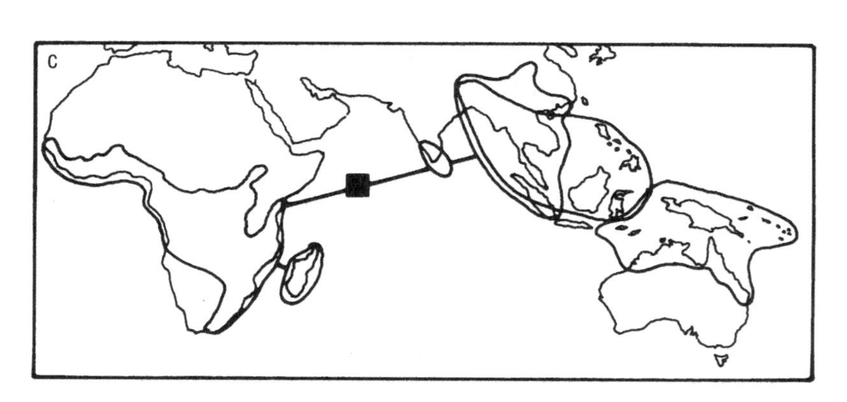

FIGURE 3-4. Classic Indian Ocean (Gondwanic) patterns in bird distributions. (a) Falconidae, *Polihierax* (three species) (distribution data from del Hoyo et al. 1994); (b) Apopidae, *Rhaphidura* (distribution data from Fry et al. 1988, Sibley and Monroe 1990); (c) Accipitridae, *Aviceda* (five species) (distribution data from del Hoyo et al. 1994).

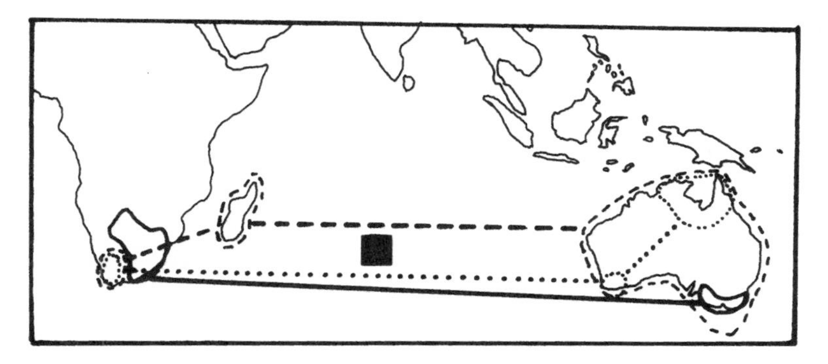

FIGURE 3-5. Vicariant patterns across the Indian Ocean for (dotted lines) *Roridula* (South Africa, Roridulaceae) and *Byblis* (Australia, Byblidaceae) (from Hutchinson 1946, Carlquist 1976a,b), (solid lines) *Papaver aculeatum* (Papaveraceae), and (dashed lines) *Caesia* (Liliaceae) (from Hutchinson 1946, Carlquist 1976a,b). Note the contrasting geographic patterns in Africa and Australia, respectively, despite the common biogeographic homology with respect to the Indian Ocean basin.

docarpus), and owls may range widely in the Indian Ocean, and yet they fail to occur on the African mainland at all. The Madagascar owl *Otus rutilus*, for example, is also represented by subspecies on Seychelles, Comoros, and Pemba islands (Burton 1973), and the flying fox fruit bats *Pteropus* (fig. 3-6c) range through much of the western Pacific, Australasia, Southeast Asia, and the Indian Ocean islands with different species on Pemba, Zanzibar, Mafia, and Comoros, but none on the African mainland (Kingdon 1974, Mickleburgh et al. 1992).

Atlantic Ocean Baselines

Geologists regularly publish geological maps of Brazil–West Africa joined together, showing in detail the continuation of geological features between the continents (e.g., Popoff 1988). Transatlantic affinities between the floras and faunas of America and Africa are also well documented. Many typical tracks connect West Africa with Brazil, the Guianas, and/or the West Indies, such as those of *Erismadelphus* (Vochysiaceae) (fig. 3-7a) and *Pitcairnia* (Bromeliaceae) (fig. 3-7b). In the Euphorbiaceae, Atlantic connections between northeastern Brazil and Gabon are shown by *Pogonophora letouzeyi* and *Conceveiba microstachys*, the only members of their respective genera outside their localized American ranges (Breteler 1993). In recognition of the Atlantic influence on African ecology, Letouzey (1966) described the *Sacoglottis* forest around the Bight of Biafra as an "archaic forest with South American affinities."

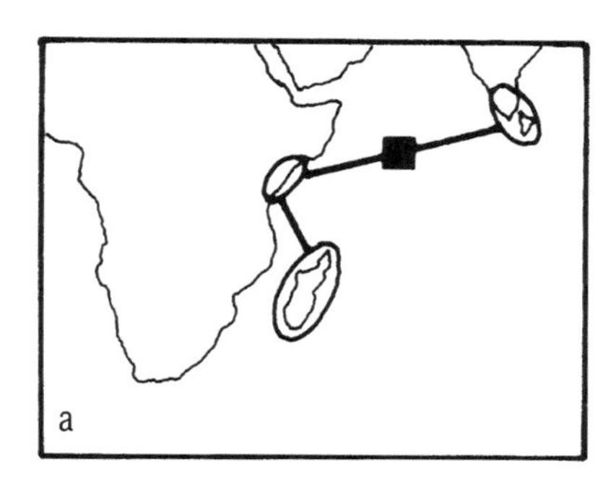

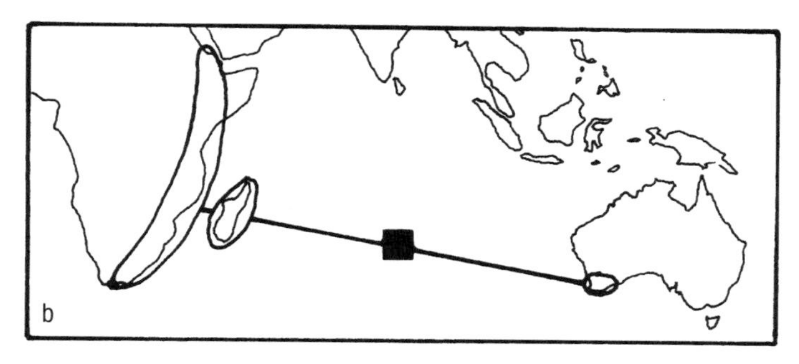

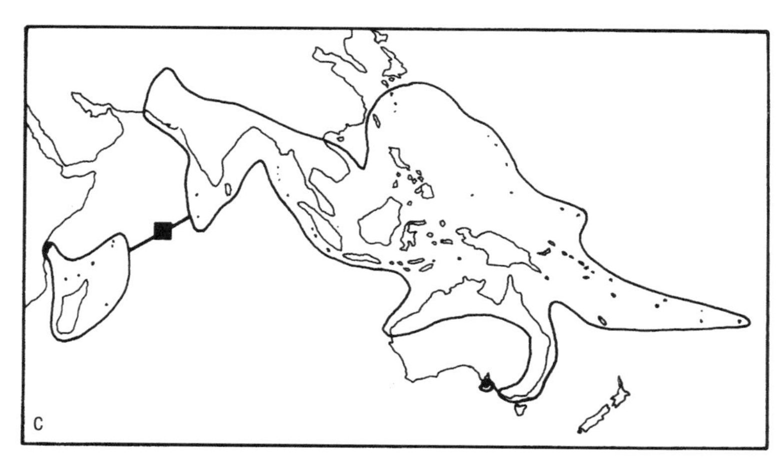

FIGURE 3-6. Examples of Indian Ocean distribution patterns that are marginal to Africa: (a) *Givotia* (Euphorbiaceae), (b) *Athrixia* (Compositae) (from Hutchinson 1946), (c) *Pteropus* bats (from Mickleburgh et al. 1992). Note how the daisy *Athrixia* is present in eastern Africa and Madagascar and the southwestern corner of Western Australia, and, although the bats are far more widespread in general, their presence in Africa is confined to two or three offshore islands. This geographic incongruence is biogeographically congruent if viewed in the context of ancestral dispersal across paleogeographic landscapes involving the tectonic basin now represented by the Indian Ocean.

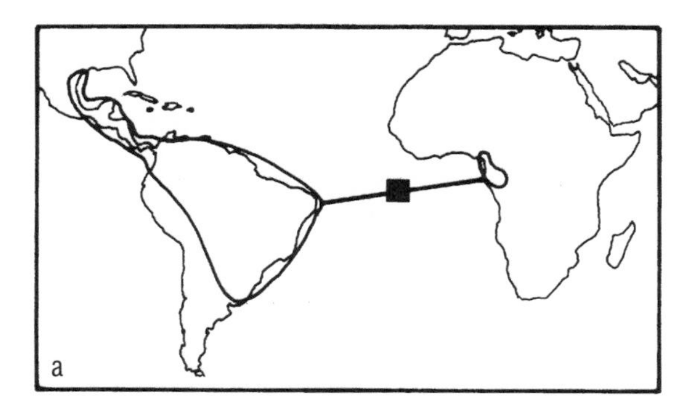

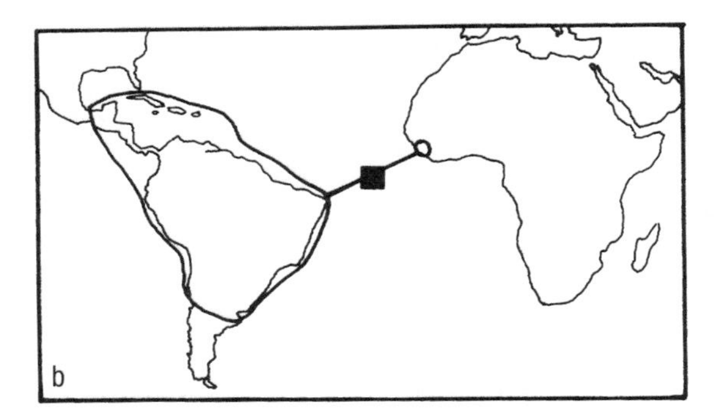

FIGURE 3-7. Atlantic baselines and generalized distributions of (a) *Erismadelphus* (Vochysiaceae) (from Keay and Stafleu 1952, Stafleu 1954) and (b) *Pitcairnia* (Bromeliaceae) (from Keay and Stafleu 1952, Stafleu 1954, Aubréville 1974b). Each distribution represents a well-known pattern where diverse and widespread taxa in the Americas are only marginally present in Africa in both diversity and distribution range.

Hystricognath rodents in the New World and Old World were long regarded by mammalian and vertebrate specialists as the result of parallelism or convergence (e.g., Simpson 1945, Wood 1950, Udvardy 1969), but their distribution pattern suggests an Atlantic baseline (fig. 3-8) and a close phylogenetic relationship (Croizat 1958, 1971). More recent studies of morphology (Woods 1982), immunology (Sarich and Cronin 1980), and parasites (Hugot 1982) confirm a monophyletic status for Hystricognatha, a position also supported by the mammologists George (1993) and Martin (1994, for Caviomorphs). The biogeographic and phylogenetic corroboration suggests that the Hystricognatha was present in Africa and America before the formation of the Atlantic, well before the earliest fossil records (cf. Wood 1980, Woods 1982). Modern placental mammals are treated traditionally as a purely post-Cretaceous phenomenon (e.g., Cox and Moore 1993), but molecular clock es-

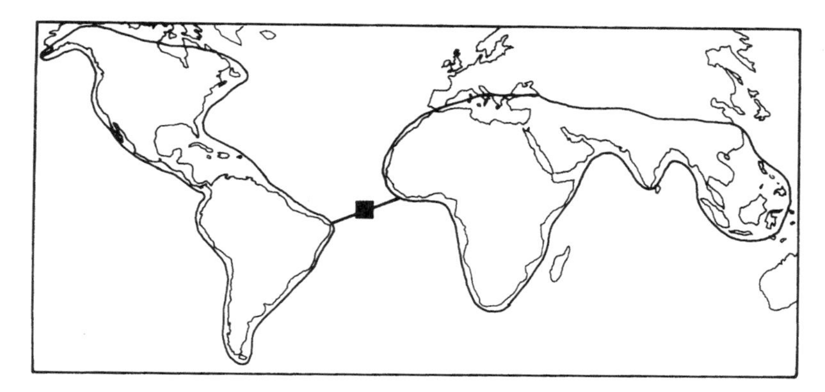

FIGURE 3-8. Distribution and Atlantic Ocean baseline of Hystricognath rodents (from Woods 1984). The main massings of species and families in South America and Africa, along with geography and phylogeny, suggest that the group was present in Africa and America before the formation of the Atlantic Ocean rather than being initially derived from either. This history may hold for many other mammalian taxa present in both continents (see text for further details).

timates for the divergence of lineages such as rodents, carnivores, ungulates, and primates range from 41 to 114 MYBP (Janke et al. 1994). Recent molecular estimates for divergence between sloths, armadillos, and anteaters range from 73 to 88 MYBP, well into the Cretaceous (Höss et al. 1996).

Indian and Atlantic Baselines

Distribution of *Thespesia* (Malvaceae) (Fosberg and Sachet 1972) illustrates both Indian and Atlantic Ocean patterns (fig. 3-9a), as well as the biogeographic importance of certain islands on the margins of Africa constituting key biogeographic nodes: Fernando Po Island (now Bioko) in the Gulf of Guinea, and Aldabra Island (north of Madagascar). *Thespesia populnea* ranges in central America (Florida, West Indies, Colombia, Guyana), and from India/Sri Lanka to Hawaii/Marquesas. Although described as "pantropical, especially on sea coasts" (Fosberg and Sachet 1972:00) it is notably neither in Africa nor Madagascar, but occurs on small islands north of Madagascar: Maldives, Aldabra, and Amirantes. *Thespesia populneoides* is found in East Africa (Kenya, Mozambique, Madagascar, etc.), has additional records at Fernando Po, and ranges from India/Sri Lanka to Queensland and Papua New Guinea (Milne Bay).

These two species of *Thespesia* are both centered on the Indian Ocean,

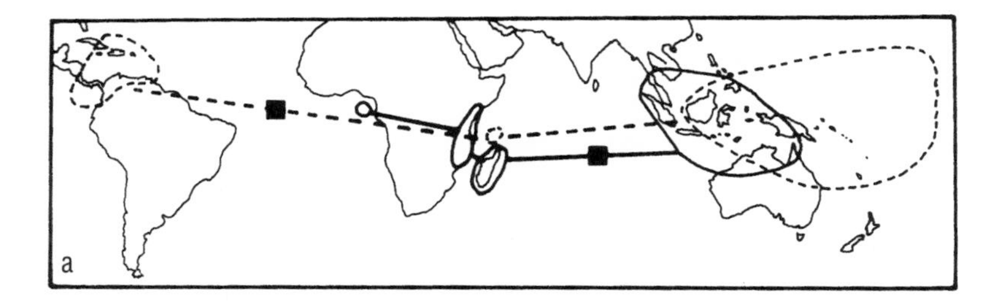

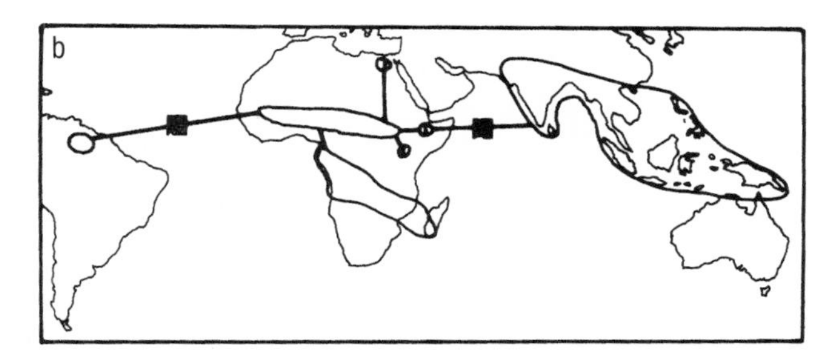

FIGURE 3-9. Atlantic and Indian Ocean baselines. (a) distribution of *Thespesia* (Malvaceae) (from Fosberg and Sachet 1972) illustrating both Indian and Atlantic Ocean patterns as well as the biogeographic importance of small islands on the margins of continents; (b) Dipterocarpaceae illustrating massings of New World representatives in northeastern regions of South America that have African affinities (from Aubréville 1974b, Londoño et al. 1995).

but in fact show a large measure of vicariance around the southwest Indian Ocean node ("African gate") and illustrate two phases of Indian Ocean dispersal, one scarcely involving Africa at all, the other heavily infesting the east coast and its hinterland. The Fernando Po record occurs at a standard Atlantic Ocean node, acting in this case as a connecting station for the genus on the Madagascar–West Indies disjunction. (The Caribbean population of *T. populnea* could be connected with the Hawaii–Marquesas population, the Aldabra population, or both.)

More primarily "pantropical" distributions are those that occur in the Americas, Africa, and Southeast Asia. Most of these taxa are known, or are likely, to have Atlantic/Indian Ocean baselines where the intercontinental relationships are with Africa rather than primarily between the Americas and the Western Pacific. The Atlantic/Indian Ocean connection may also be inferred from the massings of New World representatives in the northeastern regions of South America, as occurs in the Dipterocarpaceae (fig. 3-9b), which are limited in South America to two known species in Guyana and Amazonian Colombia, respectively, that show African affinities (Londoño et al. 1995).

3.3 African Biogeography and Ecological Lag

A particular site, as a geographic locus, will experience through time changes in elevation, slope, climate, and other environmental factors. The animals and plants present will change as well, depending on the rigors of the environment, from forest to woodland, shrubland, grassland, and desert communities. Despite these changes, the composition of the communities will be drawn from pools of species that are sometimes local in extent. Even in the densest forests there are smaller open areas, often with distinctly endemic species. These form the nuclei of species pools that can expand at any time into the forest zone, for example, during the next cycle of climate change. Communities do not really move, but there is a switching on and off of either the high-stress or low-stress biota already present. Thus biogeography often equals, or at least implies, the ecology of the past, as, for example, when formerly coastal animals and plants are stranded inland after marine regressions. This can be described as ecological lag, and ecological correlations are primarily with a past and not the present environment. The term "ecological lag" describes the tendency for ecological communities to survive, literally to live on, where they are while the physical environment changes.

What was the ecology of the Mesozoic groups, which became modernized in and around Africa, and how has this changed, if at all? In attempting to answer these questions, consider *Acrostichum aureum*, a distinctive fern found throughout the tropics. It grows typically in mangrove forests and reed beds and is able to tolerate high salinity. In southern Africa *A. aureum* is recorded along the coasts of Natal and Mozambique, and also next to hot springs in southeast Zimbabwe at an elevation of 550 m and 400 km from the coast (fig. 3-10a). This area formerly occupied a coastal and marine situation and *Acrostichum* once grew along this "inland" coastline, but as the sea receded eastward, a relic population was able to remain intact alongside a mineral spring (Burrows 1990). A similar, well-known example is the marine jellyfish found in a deep hole on the Tibesti mountains in the Sahara, a site of marine transgression during the Mesozoic (see Furon 1963, Katz 1987). Using a similar concept of ecological lag, Lonnberg (1929, cited in Grubb 1978) argued that some of the savannah mammals of Africa originated from the stranding of forest species during long-term recession of the rainforest.

As in the inland stranding of animals and plants through the recession of seas, during the uplift of plateaus and mountains, some of the originally coastal and lowland taxa will survive *in situ*. Between late Cretaceous and mid-Tertiary phases of uplift, the exceedingly smooth "African" land surface was formed (King 1967). This was "a planation par excellence" and now extends over a large part of "high Africa"

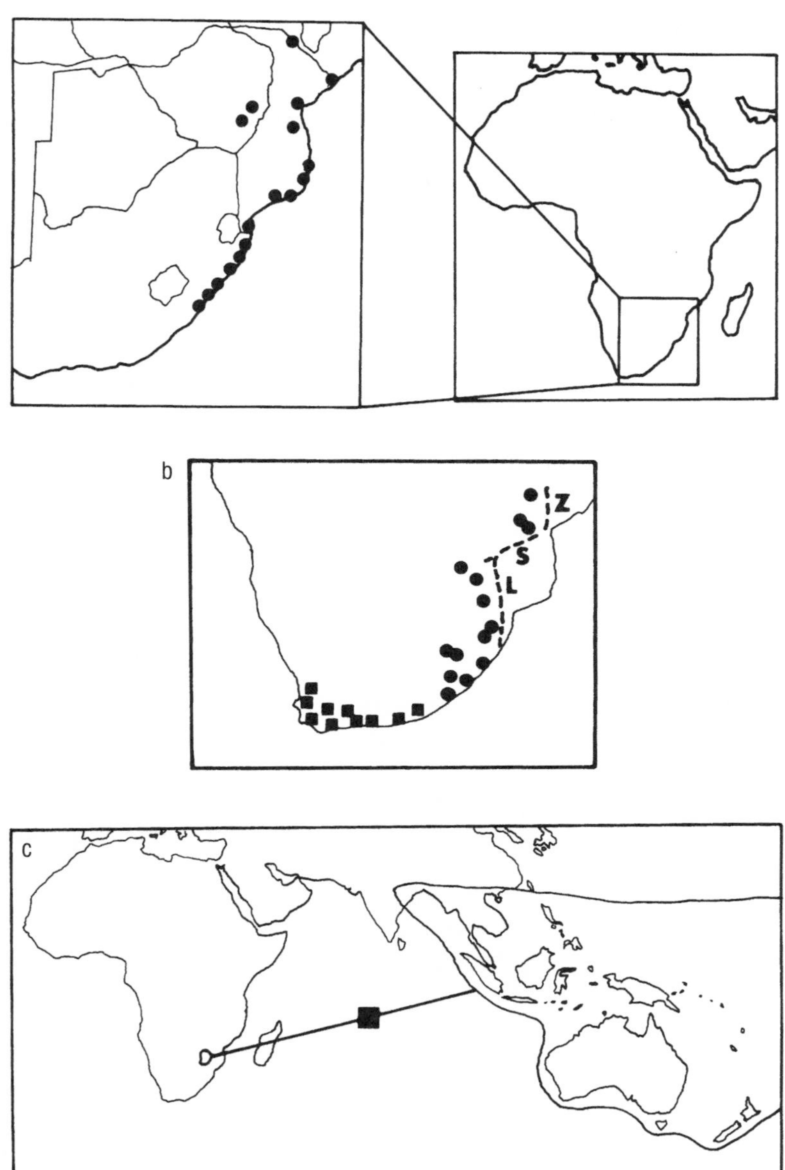

FIGURE 3-10. Examples of distributions illustrating the phenomenon of ecological lag. (a) Southern African range of the fern *Acrostichum aureum* along the coasts of Natal and Mozambique, and by hot springs in south-eastern Zimbabwe, at an altitude of 550 m; (b) distribution of avian honey eaters (*Promerops*) in Africa correlated with tectonic features associated with Mesozoic geological events. (squares) *P. cafer*; (circles) *P. gurneyi* (after Hall and Moreau 1970); (dashed line) Lebombo (L) and Sabi (S) monoclines, and Zambesi (Z) syncline (after Woolley and Garson 1970); (c) *Dianella* (Liliaceae) (after Van Balgooy 1984).

(eastern and southern Africa). King (1967) writes: "As most of it may be presumed to have stood originally only a few score or at most a few hundreds of feet above sea-level, its present elevation may be used to indicate the amount of vertical uplift that has subsequently occurred . . . ranging from 500 to 10,000 [feet]". At its highest, the land surface was raised to form the present Ruwenzori Mountains, but it was also downwarped to form, for example, the floor of Lake Tanganyka. As uplift takes place over the millennia, ancient lowland communities give rise eventually to upland and even Alpine communities. A good example is the frog *Hylambates maculatus*, well-known from tropical lowland swamps and pans in Mozambique. However, there is an anomaly in its distribution, and Poynton (1964) noted that "the occurrence of this essentially lowland form at Leopard Rock, Vumba Mountain [Zimbabwe] (4,600 [feet] = 1,400 m), a record based on five specimens, is the most extraordinary irregularity in the distribution of any southern African amphibian." As with *Acrostichum*, there has been uplift of a lowland form to a higher inland site through tectonic activity and survival because of morphological and metabolic preadaptation.

In Zimbabwe, Wild (1968) distinguished a lower altitude flora of basic soils (*Colophospermum*, *Acacia*, etc.) from a *Brachystegia*-dominated flora of acidic soils at higher altitude and treated these as different "phytogeographic elements." However, in some ways the former is always present in the latter—for example, in the flora of termite mounds. Termitaria in Africa have soils high in clays, nitrogen, and basic minerals and carry a distinctive flora and fauna. A high proportion of the angiosperm species found on southern African termitaria (e.g., Capparidaceae) are important components of deciduous woodland at lower altitudes (Wild 1952). A Mesozoic age for these floras would require geological uplift and stranding similar to that of *Acrostichum aureum* with the basic environment of termitaria allowing coastal elements to survive *in situ*.

The Mesozoic Lebombo Monocline along with its extensions, the Sabi Monocline and Zambesi Syncline, is a striking tectonic feature of southeast Africa (Woolley and Garson 1970), and the historical geology of this region was outlined by Stagman (1978). In southern and central Africa, the Karoo era of sedimentation began in the late Carboniferous to early Permian, during times of glaciation and glacial deposits. This was followed by freshwater flooding, giving ideal conditions for the growth of coal-forming plant life. Cold peat swamps occurred in quiet water between deltas, and good *Glossopteris* fossil floras are present in the middle Zambezi Valley. Slumping led to estuarine conditions, and climate-warming led to desert formation in the Triassic. The Karoo phase of sedimentation was brought to a close by vast fissure eruptions of mafic lavas, the Karoo basalts, in the lower Jurassic. At this time the Lebombo Monocline became active. The initial breakup of Gondwana, together with important earth history events giving rise to the Zambesi and Limpopo valleys, took place sometime in the upper Jurassic. Cretaceous sediments exposed on the Clarendon cliffs, Lundi River, Zimbabwe, represent a shallow-water deltaic or littoral facies of a marine incursion from the east.

Notable biogeographic landmarks at the Lebombo Monocline, at the Zambezi and Limpopo valleys, and at many other locations around the Indian Ocean can easily be explained by a phase of biological and geological modernization in the Mesozoic. Honey eaters (Meliphagidae) range across Australia, Sulawesi, New Guinea, New Zealand, and the Pacific Islands and are represented in southern Africa by the genus *Promerops*, with one of its two vicariant species closely attuned to the Lebombo Monocline (fig. 3-10b). These birds have also been placed in the families Sturnidae (Sibley and Ahlquist 1974) and Muscicapidae (Olson and Ames 1984), so they stand out as key phylogenetic forms, recombining features of two other families with those of the otherwise largely western Pacific Meliphagidae. A similar botanical example is represented by the lily *Dianella*, which also has a widespread trans-Indian Ocean distribution, but nevertheless ranges eastward on the African mainland only to the Lebombo Monocline (fig. 3-10c).

These effects extend beyond the margins of current Africa to include the geography of what is now the Indian Ocean. The lichen genus *Relicina*, for example, highlights the eastern boundary of the Lebombo–Madagascar node. Twenty-four species occur in three main massings: America, Asia–western Pacific, and Comoro Islands, which have a single species also present in the Americas and eastern/southeastern Asia (Hale 1975). This genus is conspicuous by its absence in Africa proper, but maintains an enigmatic presence on the Comoros. Geographically the islets are insignificant, but biogeographically they are situated at one of the most important centers in the world. This pattern is consistent with vicariant evolution taking place around nodes involving the Mozambique/Lebombo monoclines, which were tectonically, physiographically, and biologically active during the Mesozoic.

This line of reasoning, invoking ecological lag, follows that of Croizat (1952, 1968a,b), who argued that living terrestrial groups are descendants of Mesozoic weeds of the shore and its hinterland. The main massings of biodiversity have remained basically *in situ* since the last major modernization of earth and life in the Mesozoic. In this era there were important phases of marine incursion. During these, coastal weedy groups, such as ancestral angiosperms, spread widely within what would become Africa and Australia, as well as along the rifting zones that divided Gondwana into the present southern continents. Later, as in the case of *Acrostichum*, seas receded while living communities remained where they were, but often on new landscapes and with new ecological relationships.

3.4 Ecoclines and Ecophyletic Series

Closely related taxa often occupy different points along ecological clines, forming ecophyletic series (Aubréville 1971). Some major ecological parameters, describing what can be the determinate factors in shaping distribution, are represented by the following clines or axes: (1) light—open, high-light to closed, low-light vegetation, (2) humidity and water gradients, (3) pH— saline/alkaline to acid environments, (4) substrate particle size, and (5) elevation—coastal, inland, montane, Alpine. It is the explanation of these series, rather than the ecology of any particular habitat, that is the crucial biogeographic problem.

In many groups that are Mesozoic in origin, the ecophyletic series could be of Mesozoic age or even earlier, at least with respect to clines 1–4 above. Consider a hypothetical Mesozoic ancestral gymnospermous complex and associated phytophagous insects, with species of both coastal lagoons and sea cliffs that may have been modernized to give an extant assemblage of angiosperms with similar ecoclines to the ancestral complexes. Cline 5, however, is a function of tectonic uplift that may have been much more recent. In this case, natural selection will "prune off" forms that are not preadapted to windy, montane heights, such as plants and insects with large membranous structures.

Case study 1: the Miombo woodland

Woodland is structurally intermediate between savanna and forest in terms of grass cover and canopy closure. "Miombo" is a term given to a type of woodland that has a vast distribution in southern Africa, and like many African rainforests, it is dominated by trees in the tribe Amherstieae (Leguminosae: Caesalpinoideae). There is no full shade, and varying amounts of grasses are present. Leaf emergence precedes onset of the rainy season and often coincides with flowering. Characteristic structural features include low trees (9–12 m high) with deciduous habit; thick, often deeply furrowed bark (e.g., *Strychnos*) or fibrous bark (e.g., *Brachystegia*), and often large stipules. A gnarled habit is common, and the almost twining, horizontal stems of *Julbernardia* are conspicuous. Larger woody plants often belong to genera (such as *Terminalia* and *Diospyros*) that are more diverse in the wet tropics. These trees are often in fact giant shrubs, with several main axes and pronounced coppicing tendencies. Miniature trees composing a single, small main trunk are common and include many Rubiaceae, as in tropical rainforest.

The miombo is related structurally to different types of vegetation outside Africa. Tree architecture is often characterized by originally horizontal branch complexes that are held flat at ground level until secondary growth raises them 40–80° by the end of the rainy season (e.g.,

Acacia, Brachystegia, Combretum). In contrast, members of *Strychnos* have a divaricating growth form with the main axes vertical, and then curving by primary growth to a horizontal orientation. Other divaricating plants include *Commiphora*, which shows apical abortion (forming branch spines) and leaves borne in short shoots (brachyblasts). Miombo species of *Erythrina* also show right-angle branching and more or less distinct brachyblasts. The divaricating growth form is also a prominent feature in some Madagascar and New Zealand forests.

Lianes are generally absent, but occur occasionally on termitaria. Another distinct miombo life form is represented by low shrubs with large, woody underground stems which resprout after fire. These shrubs belong to genera that also include larger trees (*Combretum*, Combretaceae; *Parinari*, Chrysobalanaceae; *Syzygium*, Myrtaceae; *Annona*, Annonaceae; *Lannea*, Anacardiaceae). The shrubs are common on the edges of seasonally inundated grassed pans ("dambos" or "vleis"). Grasses are another rhizomatous group of plants often co-dominant in the miombo. All these life forms are also found in coastal communities, and miombo can be seen as a subset of these, while the tree architecture and growth form is reminiscent of low oak forests of central Europe (Walter 1971). The varied environments in the miombo include rocks (with *Ficus*, etc.), rivers (with riverine "fringing" forest with *Syzygium*, very like lowland rainforest), and termitaria. Soils of termitaria have higher pH, due to calcium carbonate deposits, and more water retention than the surrounding acidic, sandy soils. Wild (1952) reported that the termitaria preserve a flora that is often known only at lower altitudes.

The dominant tribe in miombo, the Amherstieae, includes many mangrove forms, and this relates to the existence of important Gondwanic boundaries and centers (i.e., nodes) at, for example, the Zambezi and Limpopo valleys. In Africa the Amherstieae are diverse through most environments, and while equally conspicuous in forest and savannah, are abundant southward only to Zimbabwe. Outside Africa, the caesalpinioid legumes (tribe Amherstieae) are abundant in forests of the Amazon, India, Sri Lanka, and Burma. Although dominant in the miombo, they are more diverse in regions such as the Gulf of Guinea rainforests where the trees are usually tall, with straight trunks and buttresses. They occur in swamp forest as well as on dry land, and occasionally form large, monospecific stands. Many other phylogenetic relations exist between the biota of the miombo and that of mangrove. *Annona palustris* is common in tidal estuaries and mangrove swamps all over South America, while other species of the genus compose shrubby growth forms in the miombo. The plagiotropic (horizontal) architecture of the shrubs with subterranean stems occurring in seasonally inundated habitats may be related to the plagiotropic habit of many mangroves. Mangroves are periodically inundated on daily and monthly cycles, whereas in the miombo the cycle is annual. In fact, Beard (1953) defined savannah as the natural vegetation on mature soils of ancient land formations with poor drainage—those exposed to temporary waterlogging at one time and extreme drying out at another. These standard affinities of miombo plants are all typically Gondwanic patterns. This suggests the vicariant differentiation of a previously widespread common ancestral complex. In this context the miombo evolved as a stranded, uplifted community of mangroves and mangrove associates.

Case Study 2: West African Rainforest

Tropical rainforest is the most diverse and least documented of all major terrestrial communities. Understanding the composition and functioning of these forests is obviously a crucial challenge for biology. Lowland tropical rainforests are distinctive for their immense trees, but there are many floristic differences on a local scale, and many of the structural peculiarities do not readily permit generalizations (Walter 1971).

There are few ecological features not found in other types of forests and woodland, and the ecological limits between rainforest and drier types of woodland are obscure. Humid tropics are often far from being the moist plant environment usually envisioned. Xeromorphic leaves are common in rainforests and enable trees to survive the short but severe dry periods occurring in all tropical areas. There is probably no lowland tropical area without deciduous trees. They are inconspicuous in the humid tropics, but become predominant in areas with more pronounced dry seasons. It is seldom appreciated that closed "rainforest" occurs widely throughout the savannah zone along many of the watercourses, even very seasonal ones. Though only a few meters wide (for example, at Victoria Falls), and not shown on broad-scale maps, large numbers of species may be conserved in patches of tropical riparian forest within savannah (Kellman and Tackaberry 1993). Lacking any single consistent boundary, the origin and ecology of these communities may be more appropriately accessed through biogeographic concepts rather than by vegetation type or current landscape and ecology (Nordenstam 1968). Distribution patterns for a sample of African rainforest trees and their relatives illustrate important regional interrelationships as well as the central position Africa holds in global biogeography.

The central position of Africa in tropical rainforest biogeography and ecology is illustrated by the "pantropical" Atlantic and Indian Ocean distribution of *Canarium*, the only large West African rainforest trees in the Burseraceae (fig. 3-11a). The Meliaceae, one of the most important timber tree families in Africa, includes the genus *Carapa*, which also occurs in Central America, the Caribbean, and northeast South America (fig. 3-11b). In Cote d'Ivoire *Carapa* is gregarious in rocky mountains

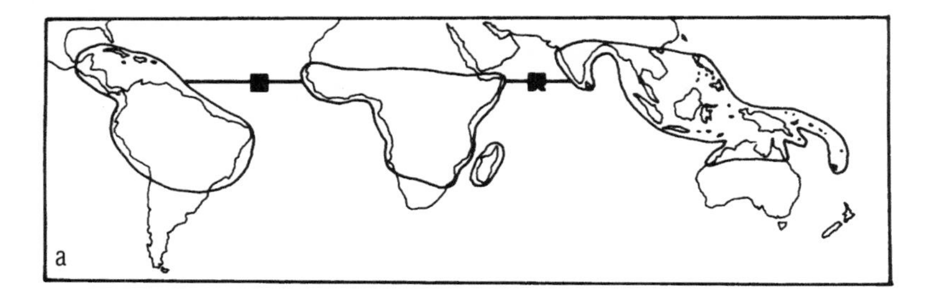

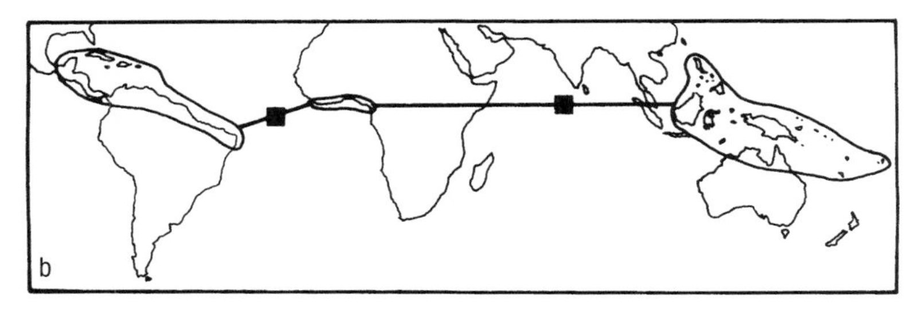

FIGURE 3-11. Atlantic and Indian Ocean baselines of African rainforest trees. (a) *Canarium* (Burseraceae) (after Van Steenis 1963, Gillett 1991), and (b) *Carapa* (Meliaceae: Carapeae) (after Pennington 1981) illustrating the importance of Indian and Atlantic ocean origins in the evolution of the African rainforest biota.

(Aubréville 1936), and in Ghana it occurs by streams in montane rainforest, with arched, occasionally rooting, branches like many mangroves (M. J. Heads, personal observation). In Malesia *C. moluccensis* is a mangrove (Airy Shaw 1973), as is the related Pacific genus *Xylocarpus* (Harms 1940), the only other member of the tribe Xylocarpeae.

Local biogeographic relationships between rainforest and other ecological communities include the littoral bush and other secondary forest formations (e.g., West African Simaroubaceae with *Hannoa* in the rainforest, and *Harrisonia* in littoral and secondary communities), and mangrove-montane ecoclines such as *Uapaca* (Euphorbiaceae) in West African rainforest. This genus includes mangroves as well as species in dry localities that still maintain aerial roots. It is widespread in tropical Africa, with species in both coastal communities and farther inland on high plateaus of eastern and southern Africa.

These affinities show a transition from mangrove to upland riverine forest and rocky mountains. A similar trend is seen in most rainforest affinities. For example, the Rhizophoraceae mainly comprises arrant mangroves, but also includes the large rainforest trees *Anopyxis* in West Africa and *Anisophyllea* (sometimes treated as its own family), which form mid-sized trees that are particularly abundant on the Fouta Djallon plateau. The family Combretaceae is a major component of African savanna and woodland, but also includes *Conocarpus* and *Laguncularia* in mangroves and saline swamps.

3.5 African Biodiversity

The African biota has been portrayed by many authors as depauperate in biodiversity. There are some strange absences. For example, *Suriana* (Simaroubaceae) is pantropical on calcareous and rocky beaches but is absent only on the beaches of West Africa (Nooteboom 1962). The American and Malesian tropics have overall more species diversity than Africa, but this species-centric view distorts understanding of global biodiversity. In the tree family Myrtaceae, many species occur in wet and dry areas of both Americas and Malesia/Australia, but remarkably few occur in Africa. This lack or absence of groups like Myrtaceae from Africa is often explained by extinction due to Tertiary aridity on the continent. However, there is no ecological reason for the absence of Myrtaceae in Africa. *Eucalyptus* is diverse in arid Australia, and *Eugenia* is in drier parts of Brazil. The comparative absence of Myrtaceae in Africa is also only apparent as it is represented in Africa by *Heteropyxis*, a genus of two species, often assigned to its own family. It possesses the aroma and other characters of Myrtaceae and may be best treated as the most distinct genus within the Myrtaceae that represents, at a high phylogenetic level, as much biodiversity as, say, the 500 species of *Eucalyptus* in Australasia.

Main massings of the monocotyledons also furnish interesting examples of absence from Africa. Palms and orchids are more diverse in both the Americas and Malesia, with many more species than in Africa. However, petaloid monocots, grasses, Restionaceae, and other groups are more diverse, in the higher taxonomic ranks at least, in Africa than in the Americas or Malesia. Again, the idea that Africa is depauperate in biodiversity comes from concentrating on species numbers and ignoring vicariance at the higher taxonomic levels that represent phylogenetic biodiversity. A further illustrative case is the genus *Rourea* (Connaraceae), which comprises three subgenera: *Jaunea* with 21 species in Africa and 1 each in Madagascar, Vietnam and Sumatra respectively; *Rourea* with 29 species in Central and South America and 1 in West Africa; and *Palliatus* with a section of 30 species in Southeast Asia, Malesia, Northeast Australia and Melanesia, and a section of 8 species, 5 in West Africa, 1 in Madagascar, 1 in the Deccan and Malay Peninsulas, and one in Borneo (Leenhouts 1958). Only Africa has all three subgenera, even though it has fewer total species than anywhere else within the overall distributional range of that genus.

The biota of the southwest Cape region of South Africa is famous for its high diversity. Many of the biogeographic relationships of organisms in this area involve South America and Australasia as well as other Gondwanic patterns. A Mesozoic origin for the Cape biota would not only account for the Australasian connection, but also for the biogeographic patterns within the southwest Cape province that correspond to the Cedarberg and Zwartberg Fold Belts which were active in the Mesozoic. These belts also intersect at the major center of biological diversity in the southwest Cape, near Cape Town.

3.6 Conclusions

Biogeography cannot be predicted from ecology alone (including means of dispersal). Ecological constraints may prevent a taxon surviving in a certain region, but they do not explain why and how the taxon was there to begin with. In fact, the ecology of a group, its relationship with its environment, is largely determined by its location—that is, its biogeography. From these considerations it can be concluded that the age of a community's present habitat does not necessarily equal the age of the community. Often a community will survive more or less *in situ* while the habitat changes under its roots and feet. Local ecology does not determine biogeography, but rather the reverse. Many African taxa are found in quite different communities, for example in both savannah and forest, but are not found throughout Africa in either. English authors have referred to these taxa as ecological "transgressors" as they "sin" by crossing established boundaries, with the subtle implication that the taxa began in one region and invaded others. French authors, on the other hand, call the same taxa *"espèces de liaison,"* suggesting a more fundamental structural relationship between forest and savannah (Lebrun 1947).

Attempts to represent ecology and history as separate, independent scales of evolution (e.g., Udvardy 1981, Blondel 1988, Brooks 1988) fail to integrate these complex patterns of ecological and biogeographical relationships. In recognition of the integral structure of ecological and biogeographic complexes, theoretical evolutionary ecology is attempting to expand the spatial and temporal scales of ecology (Brown and Maurer 1989, Cadle and Greene 1993, Haydon et al. 1994a). Riddle and Honeycutt (1990) suggest that the explicit formulation of historical biogeographic hypotheses has an increasingly important role in current evolutionary biology, resulting not only from rethinking the connections between macro and microevolutionary phenomena, but also from the construction of molecular trees that may demonstrate a history of geographic structuring (see chapter 4, section 4). Biogeographic and phylogenetic histories of organisms may be viewed as dynamic processes where reconstruction of the past is subject to the constraints im-

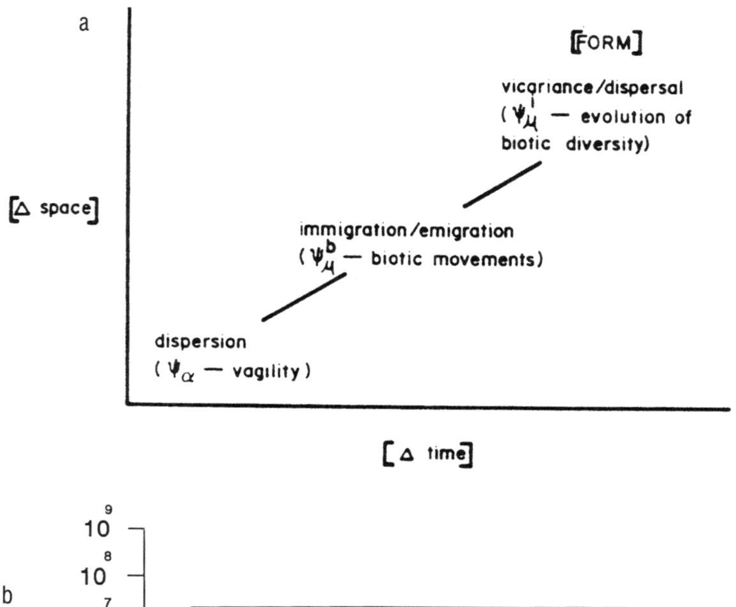

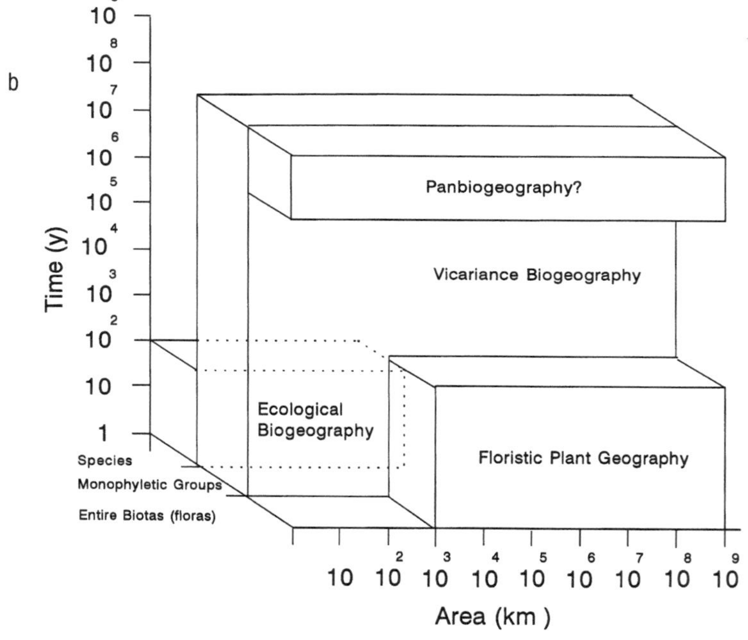

FIGURE 3-12. Examples of models that attempt to integrate spatial and temporal scales in biogeography and ecology. (a) simple linear scale separating long-term phylogenetic processes (e.g., vicariance) from recent ecological events (e.g., dispersal) (from Brooks 1988: fig. 3, with permission of *Systematic Zoology*); (b) interlocking spatial and temporal scales (from McLaughlin 1994: fig. 1b, with permission of Edward Arnold Publishers); (c) complementary relationships between biology, history and probability (from Haydon et al. 1994: fig. 1, with permission of Elsevier Science); (d) biogeography, ecology, and epigenesis as complementary factors in a corelational triad of space, time, and form (from Craw and Page 1988: fig. 14, with permission of John Wiley and Sons).

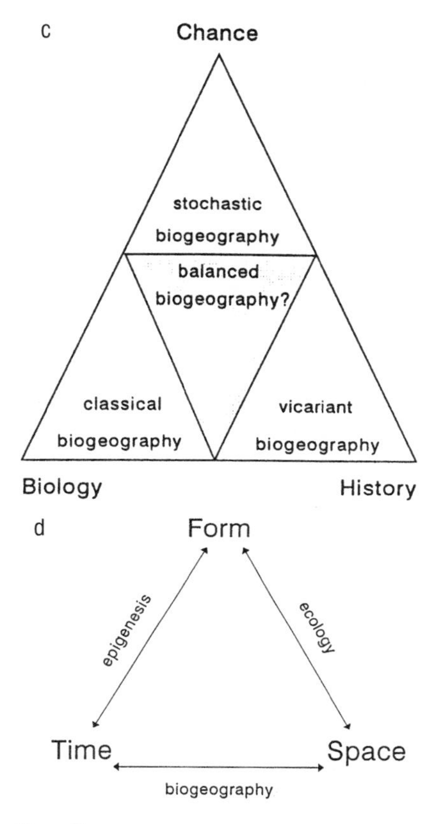

FIGURE 3-12. *(continued)*

posed by changing ecological and geographic relationships (Brown 1995).

These developments necessitate new models of biogeography as a discipline in relation to other sciences. Simple linear scales (e.g., Udvardy 1981, Brooks 1988) separating long-term historical biogeographic and phylogenetic processes from more recent ecological events (fig. 3-12a) are being replaced by more complex interlocking scales of biogeographic, ecological, and phylogenetic differentiation in space and time (e.g., McLaughlin 1994) (fig. 3-12b). Haydon et al. (1994) suggest that the biogeography of different taxa and regions may be conceptualized as a triadic structure involving complementary relationships between biology, history, and probability (fig. 3-12c). A similar dynamic interface between biogeography and other biological disciplines is represented in the panbiogeographic model of space, time, and form (fig. 3-12d). It is through the development of these conceptual frameworks that biogeography may be recognized as one of the principal foundations for ecological theory and evolution, as it was in Darwin's time.

4

Mapping the Trees of Life

Panbiogeography, Phylogenetic Systematics, and Evolutionary Processes

An understanding of geographical evidence is basic to interpreting the phylogeny, systematics, and taxonomy of species and higher taxa. The use of geographical distribution and other biogeographical data as a source of classificatory characters and phylogenetic evidence is a time-honored practice among systematists (e.g., Wettstein 1896, 1898; Diels 1924, Smith 1934, Schilder 1952, Marx and Rabb 1970, 1972; Heyer 1975, Olsen 1979, Stuessy 1979, de Jong 1980, Crisci 1980, Crisci and Stuessy 1980, Stuessy and Crisci 1984, Fisher 1994, de Queiroz et al. 1995, Lichtwardt 1995). Geographic records are an important category of evidence for problem solving in systematics (Radford 1986, Ross 1974), and critical knowledge of the type locality and the geographic status of a taxon is central to the discussion and resolution of many nomenclatural problems in the taxonomy of plants and animals (e.g., Croizat 1943 on the typication of *Echinocactus*). Correlations of distribution ranges with similarities or differences in morphological, genetical, biochemical, or other features constitute one basis for determining phylogenetic relationships at all levels of the taxonomic hierarchy.

De Candolle (1855) was one of the first systematists to note the value of geographical distribution as a guide in monographic studies of genera and families; a knowledge of geographical distribution was regarded as being particularly important for the systematics of cultivated plants (de Candolle 1886). The term "provenance" has since come into widespread use to designate the place of origin of seeds or plants, and this idea has proved vital in the study of many cultivars (Turnball and Griffin 1986). Wallace (1855) emphasized that the natural sequence of affinities of species was geographical and used this principle in his studies on Indo-Malayan swallowtail butterflies (Wallace 1860). In *On the Origin of Species*, Darwin (1859) commented on the value of using geographic evidence in taxonomic studies in both a general sense and in relation to monophyly.

Biogeographic characters are important at both micro- and macrotaxonomic levels (Mayr and Ashlock (1991), and they can be used as primary characters in the initial delineation of taxa or as secondary characters to evaluate systematic hypotheses based on other kinds of characters. Geographic data correlated with structural or other features have provided evidence for relationships of taxa at generic and family levels (Thorne 1975, 1989), while at the species level, sympatric or allopatric relationships of populations are often crucial in ranking decisions. Analysis of measured characters in the dragonfly *Tramea*, for example, revealed differences in relation to geography suggesting that the two morphological forms should be treated as separate species (Garrison 1992). In a study at and below the species level for North American copper butterflies (*Lycaena rubidus* complex), character traits of numerous populations were individually mapped and then superimposed to illustrate their geographic relationships. These "topographic maps of affinities and characters" were then used to assist in the delineation of taxa (Johnson and Balogh 1977). Distribution of populations in relation to biogeographic gradients can be combined in multivariate analyses (e.g., spatial autocorrelation) with other characters to clarify and resolve systematic and taxonomic problems (e.g., Estabrook and Gates 1984).

The panbiogeographic project builds on this widespread respect for geographic data in practical systematic studies by going one step further. It advocates that biogeographic analysis is a necessary prerequisite for a coherent representation of phylogeny and construction of a natural classification for any group of organisms (Croizat 1961, Craw 1988, Chiba 1989, Climo 1990, Smith 1990). Patterns of descent, branching, and hybridization are space- and time-dependent processes. There is an ordering of taxa across geographic space and through geological time. Panbiogeography proposes that aspatial morphological and molecular systematics be complemented by a spatial systematics and palaeontological studies.

4.1 Geographical Distribution as a Systematic Character

For biogeographic characters to be of use in the study of the classification, phylogeny, and systematics of organisms, they must be inherited. Although early systematists regarded the exact locality of a species as of little importance, by the late nineteenth century the notion that environment might be inherited was becoming accepted. Wallace (1902) insisted on the importance of locality as "an essential character of species." Being at location *x* at time *t* is an important property of any organism (Shanahan 1992). Both morphology and area are an expression of a taxon in nature and are an outcome of its history (Davis and Heywood 1963). As Shaw (1985: 202) notes: "When using differences in geographic distribution to support taxonomic separation, the implicit interpretation is that the ranges reflect different evolutionary histories for the taxa and that their different evolutionary histories imply past (if not present) genetic differentiation (especially when considered in conjunction with morphological discontinuities)."

Although many workers have recognized the value of biogeographic data for species and higher level systematics, some have rejected the use of geographic evidence in phylogenetic reconstruction and in the resolution of taxonomic problems (e.g., Borgmeir 1957, Platnick and Nelson 1989, Humphries 1990, Schindewolf 1993). This rejection appears to be based on the view that certain intrinsic features of organisms such as morphological characters are more accurate markers of history than extrinsic ecological and biogeographical variables. This dichotomy between organism and environment assumes a simple correspondence between developmental inputs and organic outcomes (fig. 4-1a) that require a separation between environmental information producing acquired characters and genetic information producing innate characters. If one source of developmental information is cut off (e.g., experience), what develops must be due to the other factors (genes). This dichotomous model overlooks the absence of a simple correspondence between DNA base-pair sequence and the functional activity of the proteins they are claimed to code for. Interaction between genes and environments is widely recognized as affecting the expression of genes, but it is genes that are perceived as the distinct and separate carriers of information (Gray 1992).

An alternative approach to genes and environments is a developmental systems or constructionist approach in which genes are recognized as just one resource that is available to the developmental process. Although particular resources may have different roles, there is nothing dividing the resources into fundamentally differing kinds such as genes and environment. Phenotypic traits require both genetic and environmental inputs. The fundamental unit of evolution becomes the developmental system. This is a set of organismic and environmental features interacting to produce an outcome capable of replicating the developmental process (Griffiths 1992, Griffiths and Gray 1994). Characters are constructed through epigenetic processes that are no less a codeterminant of characters than genes (Oyama 1985, Ho 1988). Evolution is the differential replication of these systems.

Numerous field and laboratory studies of ecology and development suggest that all phenotypes are the joint product of "internal" and "ex-

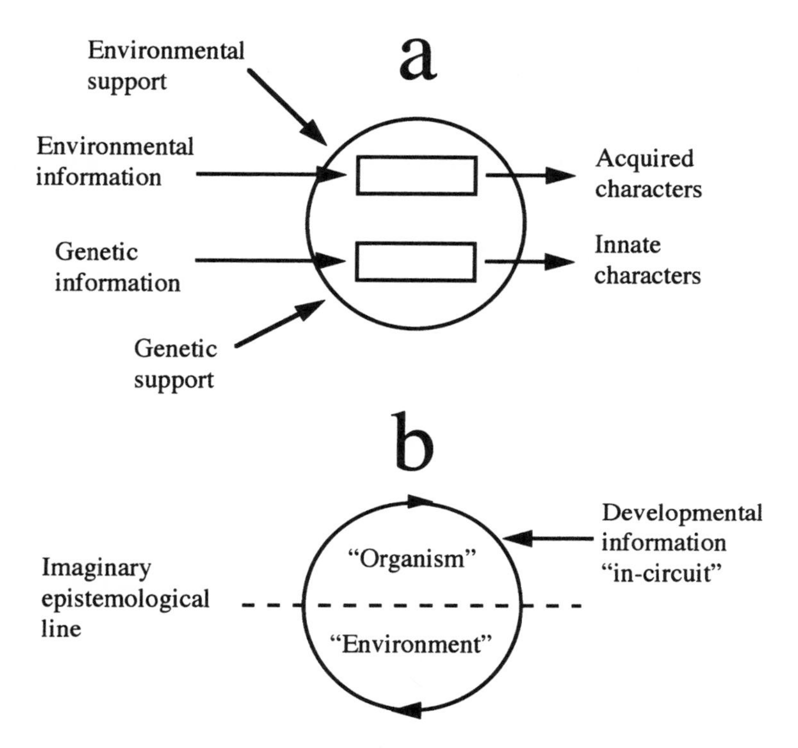

FIGURE 4-1. Organism–environment models. (a) Dichotomous view of development where distinct developmental inputs (genetic and environmental) result in two kinds of characters (acquired and innate); (b) cyclic view of development where developmental information is "in circuit" rather than being localized in either the "internal" or "external" factors (from Gray 1990: figs. 1, 2; reprinted with permission of SIR Publishing).

ternal" factors interacting through complex, nonlinear, dynamic systems. Internal and external factors represent codependent variables that do not exist in any meaningful way in isolation from each other. An organism and its environment do not exist prior to their relationship. Thus, a gene can only be functionally defined in a specific developmental context, and an organism's environment can only be identified in relation to that organism. This environment plays an essential role in the construction of its phenotype, and reciprocally, the organism selects and modifies (i.e., constructs) this environment out of the resources that are available (Gray 1992).

A constructionist view of inheritance and development leads to the concept of phenotypes being constructed each generation rather than transmitted between generations. Traits are not directly transmitted across generations, and neither are blueprints, potentials, programs, or information for the traits. Genes do not contain information for traits independently of developmental processes (Gray 1988, 1992). Developmental information required for the formation of characters or traits lies neither within genes nor environments, but in the contingent relation between them. As information is not localized in the genes or any particular entity, the control of development does not reside in the genome but is disseminated across the entire developmental system, which is constantly in circuit (fig. 4-1b).

This extended concept of heredity involves "the passing on of all developmental conditions, in whatever manner" (Oyama 1985). In-circuit inheritance includes a large set of developmental interactants encompassing cytoplasmic factors (e.g., mitochondrial DNA in animals), chemical cues, symbionts, behavioral traditions (e.g., feeding methods), physical conditions, and geographic range. Thus, even geographical distribution constitutes a suite of spatial character traits capable of phylogenetic propagation, and extra-genetic inheritance can include migration routes, schooling sites, home ranges, territories, and broader geographical range (Gray 1992, Jablonski 1987, Jablonski and Valentine 1990, Brown 1995).

4.2 Predicting Phylogenetic Relationships from Biogeographic Data

Geographic characters represent a source of potential falsifiers in phylogeny reconstruction and may be treated as another set of systematic characters along with those of morphology, development, genetics, behavior, ecology, and biochemistry. In this context systematic hypotheses generated from either biogeographic or other evidence may be compared and used to contribute toward an understanding of phylogeny under the principle of reciprocal clarification. This reciprocal relationship is much closer to describing the actual practice involved in many taxonomic and systematic investigations than the privileging of any other character set over that derived from geographic distribution.

Differentiation in many plant groups along ecological gradients in factors such as light, moisture, and temperature is correlated with relatively minor characters (e.g., succulence and size and hairiness of parts). Major structural changes and morphological innovations that characterize major clades and diagnose higher categories in taxonomy exhibit biogeographic rather than ecological correlation. Discussing the "geography of characters" Exell and Stace (1972) illustrated how parallelisms can be distinguished from homologous characters on the basis of geographical criteria. Knowledge of broad-scale geographic distribution patterns can contribute to the formulation of hypotheses of phylogenetic relationships by determining polarity in character-state trees, as demonstrated for the lichen family Megalosporaceae (Sippman 1983).

Reciprocal illumination between molecular systematics and biogeography was applied to an evaluation of heliornithine bird DNA trees for their reproducibility and consensus with most parsimonious biogeographic, palaeontologic, and traditional classifications (Houde 1994).

Patterson (1981) noted that the geographic distribution of a taxon may indicate that some phylogenetic histories are more likely than others. In a study of southwestern North American montane chipmunks, it was possible to predict patterns of divergence among isolated populations on the basis of biogeographic considerations (Patterson 1982). Geographic data can then be used to both determine unlikely cladogram topologies and to determine whether homoplastic characters are convergences or parallelisms (Thewissen 1992).

The Vicariance criterion

An attempt at formalizing this proposition that geography is a measure of historical descent and can be profitably used in systematics to generate novel hypotheses and test established ones, was made by Hennig (1950, 1966). He proposed that:

> The geographic distribution of organisms provides another way of checking the reliability of systematic results. Following the principle of reciprocal clarification it is possible in reverse order to use geographic distribution for determining the phylogenetic relationships themselves. In general, two taxonomic groups that stand in a spatial vicariance relationship to each other are more closely related than either is to any other taxonomic group. (Hennig 1966:148)

Stated as a general principle, this criterion postulates that spatially vicariant taxonomic subgroups (e.g., subspecies of a species or various species in a genus) may be considered as being more closely related to each other than either is to other taxonomic subgroups (e.g., another species in the same genus sympatric with either or both vicarious species) within the same group. Kiriakoff (1954, 1959, 1961, 1964, 1967, 1981) discussed this approach extensively, often in relation to panbiogeography.

Consider the five endemic species of the small, flightless, litter-inhabiting weevil genus *Etheophanus* (Coleoptera: Curculionidae) variously distributed amongst the three main islands of the New Zealand archipelago. There are three completely vicariant taxa (*E. pinguis*, *E. nitidellus*, and an undescribed species), a fourth species that is sympatric at northern and southern ends of its range with the undescribed species and *E. nitidellus*, respectively, and a fifth more widespread species (*E. striatus*) sympatric in some parts of its range with three out of the other four species (fig. 4-2a). Under the vicariance criterion, the three taxa (*E.*

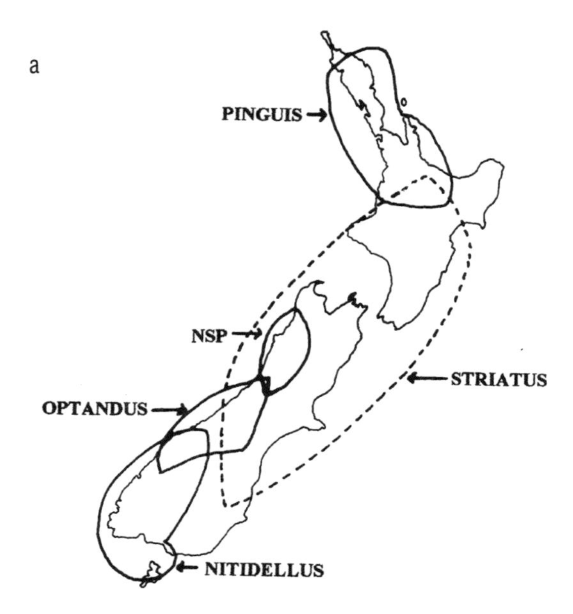

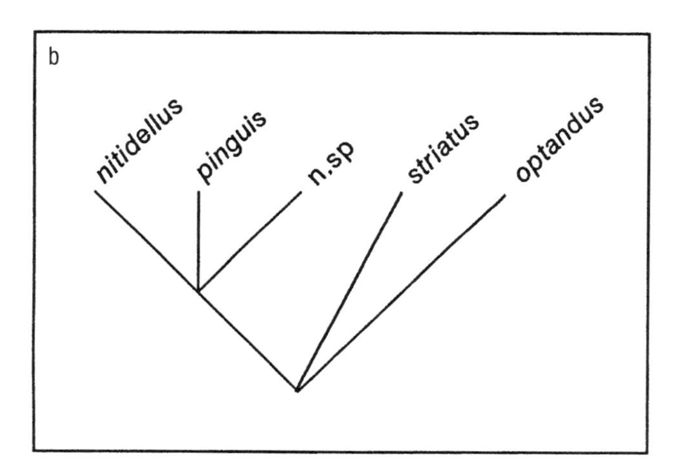

FIGURE 4-2. The vicariance criterion as illustrated by the New Zealand flightless weevil genus *Etheophanus*. (a) Distribution of the species among the three main islands of New Zealand; (b) hypothesis of phylogenetic relationships based on these distributions (from Craw 1989: fig. 3; reprinted with permission of *Systematic Biology*).

pinguis, E. nitidellus, and the undescribed species) that stand in a spatial vicariant relationship to one another can be regarded as being more closely related to one another than any of them are to the two other *Etheophanus* species (Craw 1989). A hypothesis that these three species form a monophyletic clade relative to the other two species can thus be proposed on the basis of their biogeographic relationships (fig. 4-2b).

A further example involves possible relationships among subspecies of the Australian rat *Rattus fuscipes*. There are four subspecies, each endemic to a different part of Australia (fig. 4-3a). Given this geographic distribution, certain hypotheses of relationship are considered to be more likely than others (fig. 4-3b) (Page 1991). Applying the vicariance criterion clearly introduces an independent measure of phylogenetic relationship that is testable and recognizes the integral contribution of biogeography to systematics.

Case Study 1: Vicarious moths

Larvae of the ghost moths (Lepidoptera: Hepialidae) are mostly root or ground feeders living in subterranean tunnels, but a number of genera are characterized by larvae that are arboreal wood-borers in trees and shrubs. This feeding behavior is largely restricted to genera distributed in Central and South America, eastern and southeastern Asia, and the southwest Pacific. Asian and southwest Pacific species comprise two vicariant groups, *Endoclita* and the closely related *Sahyadrassus* to the west of the Moluccas and Banda Sea, and *Aenetus* and a close relative (*Zelotypia*) to the south and east (fig. 4-4). Under the vicariance criterion, this distribution suggests that the two massings are the result of a phylogenetic affinity involving a former widespread ancestor. The Moluccas and Banda Sea region includes the Flores/Soemba and Molucca nodes, which characterize many major Indo-Australian distribution patterns (Croizat 1958), supporting an evolutionary model involving differentiation of a widespread ancestor in concert with biogeographic events affecting the general regional biota (Grehan 1987).

When the biogeographic hypothesis was first considered, a close phylogenetic relationship between endoclitine and aenetinine genera was considered unlikely by hepialid systematists. Subsequent morphological studies have now shown that the two groups share characters, particularly larval and pupal ones, that are suggestive of a close relationship relative to other hepialid genera (Dugdale 1994). Although the details of this relationship are subject to future analysis, there is a correspondence of spatial vicariant relationships and morphological characters. If future systematic studies of the ghost moths result in several topologically conflicting yet equally parsimonious trees for these genera

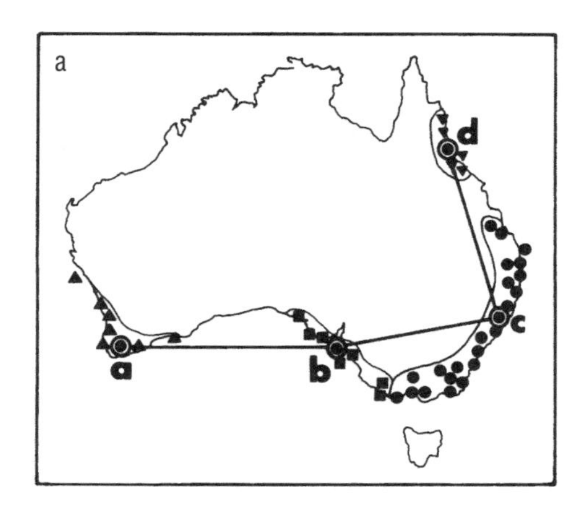

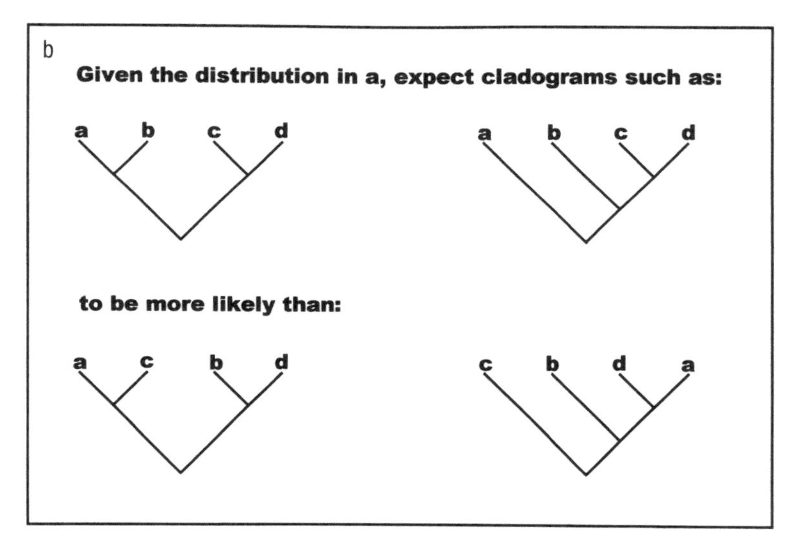

FIGURE 4-3. The vicariance criterion as illustrated by the subspecies of an Australian rat species. (a) Distribution and track of *Rattus fuscipes* subspecies. Track drawn between centroids of distribution for each subspecies; (b) likely and unlikely cladistic relationships of these subspecies as constrained by geographic distribution (from Page 1991: fig. 4; reprinted with permission of *Systematic Biology*).

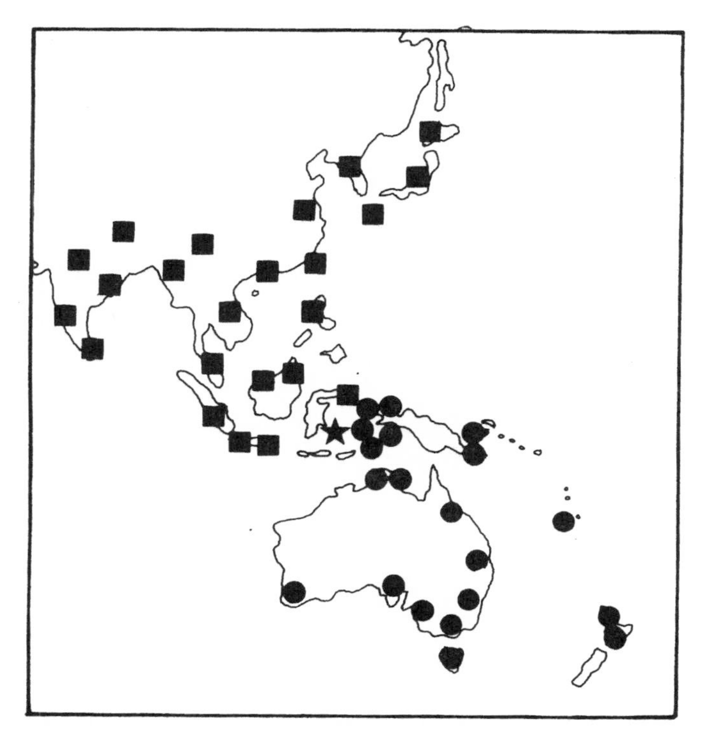

FIGURE 4-4. Vicarious distributions of *Endoclita/Sahyadrassus* (squares) and *Aenetus/Zelotypia* (circles). Banda Sea located at star (from Grehan 1987: fig. 20; reprinted with permission of SIR Publishing).

(fig. 4-5), some sets of trees may be considered to be more likely (fig. 4-5a, b) than others (fig. 4-5c, d) under the vicariance criterion.

Case Study 2: figs and wasps

The widespread distribution of the figs (*Ficus*, Moraceae) in both the Old World and New World illustrates problems with biogeographic homology because all three major ocean basins (Atlantic, Indian, Pacific) are implicated as possible baselines for the distribution of the genus. Characters of Melanesian, Micronesian, and Polynesian *Ficus* (sections *Urostigma* and *Pharmacosycea*) are shared with American species in the same sections and have been taken as evidence of a direct phylogenetic affinity. These similarities have been explained by migration across a transpacific land bridge (Corner 1963) or across the continent of Pacifica (Corner 1985).

A combined morphological and biogeographic study suggested that

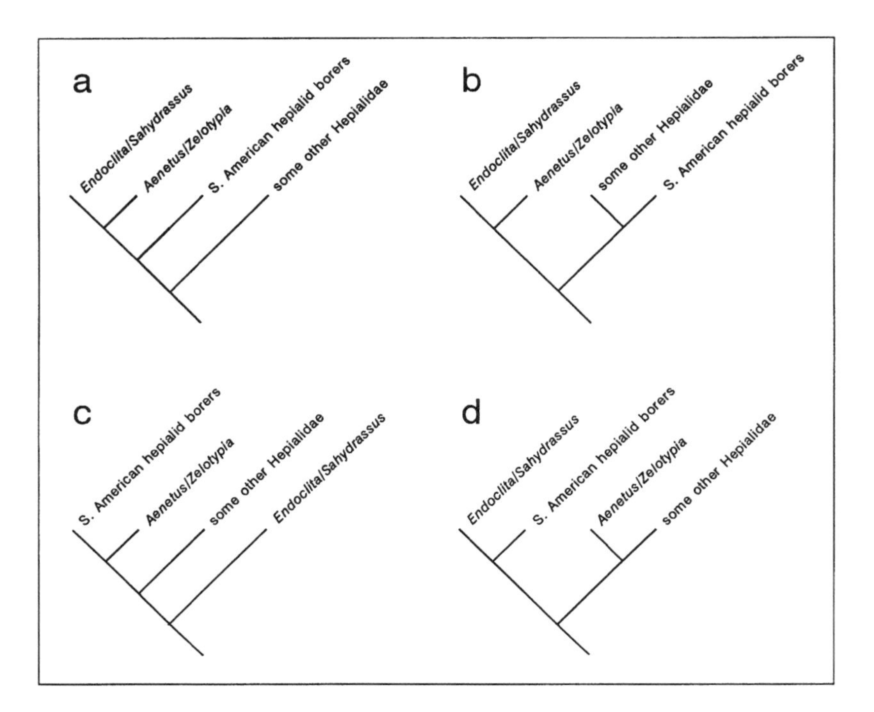

FIGURE 4-5. Some alternative possible cladogram topologies for *Endoclita, Aenetus*, and other wood-boring Hepialidae. (a) and (b) Predicted relationships congruent with the vicariance criterion; (c) and (d) relationships not predicted by the vicariance criterion.

sections *Urostigma* and *Pharmacosycea* figs were not primarily Pacific, but rather centered on the Indian Ocean (Croizat 1968a). In particular, the western and central Pacific species *Ficus prolixa* was shown to be an eastern vicariant outlier of a closely related group of figs distributed through the Indo-Malaysian region to eastern and southern Africa (fig. 4-6). This species group as a whole has a primary Indian Ocean baseline, and *F. prolixa* represents an eastern extension of a distribution with connections to Africa and the Bay of Bengal, in common with the Moraceae in general. In this biogeographic context, similarities between New World and Pacific figs result from the recombination of ancestral characters (equivalent to shared primitive characters) without a direct transpacific relationship.

In cladistic terms, the transpacific hypothesis considers the American and western and central Pacific figs as most closely related compared with African species (fig. 4-7a), whereas the trans-Indian Ocean hypothesis suggests a closer relationship of Pacific figs with African ones (fig. 4-7b). The phylogenetic implications of the trans-Indian Ocean hypoth-

FIGURE 4-6. The vicariant biogeography of *Ficus prolixa* (stippled) and *Ficus* section *Urostigma*, series *Caulobotryae* (vertical lines), which includes the closely related species *F. virens*. The Indian Ocean range of most of these related species suggests that the western Pacific distribution of *F. prolixa* represents an eastern extension of a distribution with a primary Indian Ocean baseline (from Croizat 1968a; reprinted with permission of the Istituto di Botanica, Pavia).

esis are that neither sections *Urostigma* nor *Pharmacosycea* are natural taxonomic groups; the transpacific hypothesis implies that they are (Page 1987, 1990b).

Cladograms of taxa parasitic on a host taxon have the same logical status as characters state trees for characters of the host. Both contain information on the cladistic relationships of the host taxon, and both are a priori valid sources of information (Brooks 1981). An independent source of phylogenetic information on *Ficus* relationships is provided by a cladistic analysis of obligate symbiont agonid wasps that pollinate fig flowers. Neither of the two fig sections is pollinated by a monophyletic group of wasps. Pollinators of the Old and New World *Pharmacosycea* figs belong to different subfamilies; the precise relationships of the west Pacific and American *Urostigma* pollinators are unresolved (Wiebes 1982). This lack of concordance between the fig and wasp lin-

FIGURE 4-7. Hypotheses of cladistic relationships in *Ficus* between Africa (AF), western and central Pacific (WCP), and New World (NW). (a) Corner's hypothesis based on selected morphological characters; (b) Croizat's hypothesis based on the Indian Ocean massing of *Ficus* diversity (from Page 1987; reprinted with permission of *Systematic Biology*).

eages suggests that the proposed trans-Pacific relationships of some *Ficus* species require reappraisal. In this context, a coevolution model may provide a method of extending the set of potential falsifiers of a taxon cladogram to include distribution (Page 1987, 1990b).

4.3 Biogeography and Character Recombination

Concepts of Ancestors and Evolutionary Processes

Evolutionary biologists refer regularly to taxa at all levels as having descended from a single, common progenitor and to allied species and genera as descending from the same parent or parent form. Thus, species and other natural taxa are regarded as the descendants of initially uniform, undifferentiated, and local populations, entities, or even single individuals. A classic example is Darwin's (1859) proposal that the ancestor of a species endemic to isolated islands was either a fertile individual or a viable seed. Mayr's (1942) concept of speciation begins explicitly with uniform species that differentiate into subspecies, with subsequent differentiation of each of these into new species, and this idea is still current in biology (Heads 1985).

An alternative notion of ancestors was proposed by Rosa (1918), who maintained the species label but postulated that differentiation, as visible phenotypic differences, had already occurred before the appearance of descendant taxa. This idea of an already differentiated ancestor may strike many as unusual, but there are many references to it in contemporary literature. For instance, Wiley (1981) discusses the possibility of residual geographic variation left over from an ancestral species and notes how this can complicate phylogenetic analysis. Such complications emerged in a study of phylogenetic relationships of Old World monkeys in the genus *Macaca*, where it was noted that "speciation from a highly polymorphic ancestor can interfere with our attempts to reconstruct the historical pattern of cladogenesis" (Hoelzer et al. 1992). Implications of polymorphic ancestors with reference to the phylogeny of the plant group Lamiales were discussed by Cantino (1982). Likewise, Heslop-Harrison (1983:15), in discussing the ancestry of flowering plants noted, "we find it difficult to conceive of a situation so fluid that a single breeding population might have been heterogeneous in characters now regarded as of familial significance. Yet this may have been the starting point for the diversification, and in this fact may lie the solution to Darwin's abominable mystery."

This contrast between a uniform or a differentiated ancestor raises important questions about evolutionary processes. The evolutionary integrity of a species has been expressed traditionally in terms of cohesion being maintained by gene flow between its constituent populations, but this scenario may be unrealistic when applied to widespread taxa. Descent through modification from a widespread, differentiated ancestor does not require evolutionary novelty to evolve by chance at one place and from there spread throughout a population. This construct of a differentiated ancestor suggests cohesion is not simply a matter of gene flow, but a process of spatial and temporal ordering where other mechanisms of cohesion such as biogeographic, developmental, and phylogenetic constraints may be involved. These constraints were described by Croizat (1964) as character recombination, where during a mobilist phase characters become redistributed over geographic space, and the differential combinations of characters provide the starting points for local differentiation in immobilism.

Character Geography

In a vicariant form-making model of differentiation, vicarious taxa represent geographic and phylogenetic fragments from the ancestral range. Differentiation of descendant taxa over a widespread ancestral range may be explained as a process involving prior geographic variation in the distribution and composition of ancestral characters. Ancestral characters (a+b+c+d) are geographically redistributed or recombined in different descendants (a+b), (a+c), (b+d) (Croizat 1964, Grehan and Ainsworth 1985, Heads 1985). Because the taxa have all evolved from a widespread ancestral range, it is the ancestral range that is the common or shared center of origin for all vicarious taxa and character combinations, whether primitive or derived. It is not necessary, however, that vicarious taxa all appear at exactly at the same time. There may be a spatial and temporal sequence of differentiation, and this may be represented by the sequence of character affinities (fig. 4-8). Even though the most derived taxa may be located at a distance from the most primitive, this does not mean that the primitive taxon represents a localized center of origin from which the derived taxa migrated.

Vicariant taxa have a widespread origin in the sense that the ancestral range is broader than those of individual descendants. In this biogeographic context, characters inherited by descendants are not localized at a single point or locus, but spread over a series of points. Inheritance will be affected by the different character complements present in different geographic sectors of the ancestral range. Under these conditions it is possible for different character combinations to result from the same historical event. Conflicting gene and species phylogenies, for example, may occur when descendant taxa inherit samples of polymorphic alleles that are random with respect to the genetic lineage of those alleles (fig. 4-9). Ruvolo (1994) notes that ancestral polymorphic alleles are not necessarily randomly distributed throughout an ancestral population, but may instead show geographic "substructuring" caused

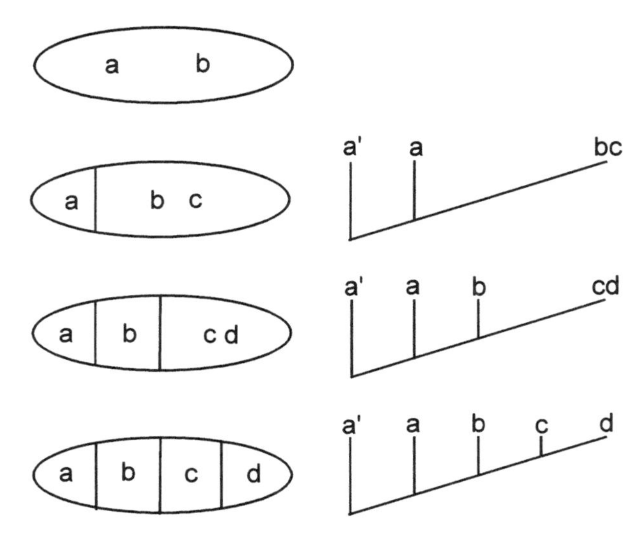

FIGURE 4-8. Spatial and temporal sequence of vicariant differentiation. (Left) Geographic distribution of characters involved with differentiation of vicarious taxa; (right) phylogenetic sequence represented by character relationships. Letters denote characters or character combinations associated with a particular differentiation event, but all may have been present as components of the ancestral range. Out-group represented as a'. In this phylogenetic sequence a monophyletic ancestor is widely distributed relative to its descendant vicariants. The individual center of origin for each taxon is represented by the different parts of the ancestral range, but the common or shared center of origin for all taxa is the entire ancestral range, even though more primitive taxa may represent earlier differentiation events.

by isolation by distance, geographic barriers, lineal fissioning, and/or social structure. Thus, characters and genes may have specific concentrations within the ancestral range, and these clusters may be correlated with different centers of differentiation.

Characters shared between different descendants that are also correlated with the geographic proximity of descendants (i.e., adjacent) may become uniquely shared by subsequent adjacent taxa and qualify as synapomorphies. Those characters more widely distributed may be inherited by several taxa and be interpreted as homoplastic (i.e., not uniquely shared), particularly if they represent less parsimonious combinations, even though both sorts of characters were inherited from the same ancestor at the same time. This process could be described as the geographical and differential fixation of ancestral polymorphisms.

Character recombination is complex and problematic when phyloge-

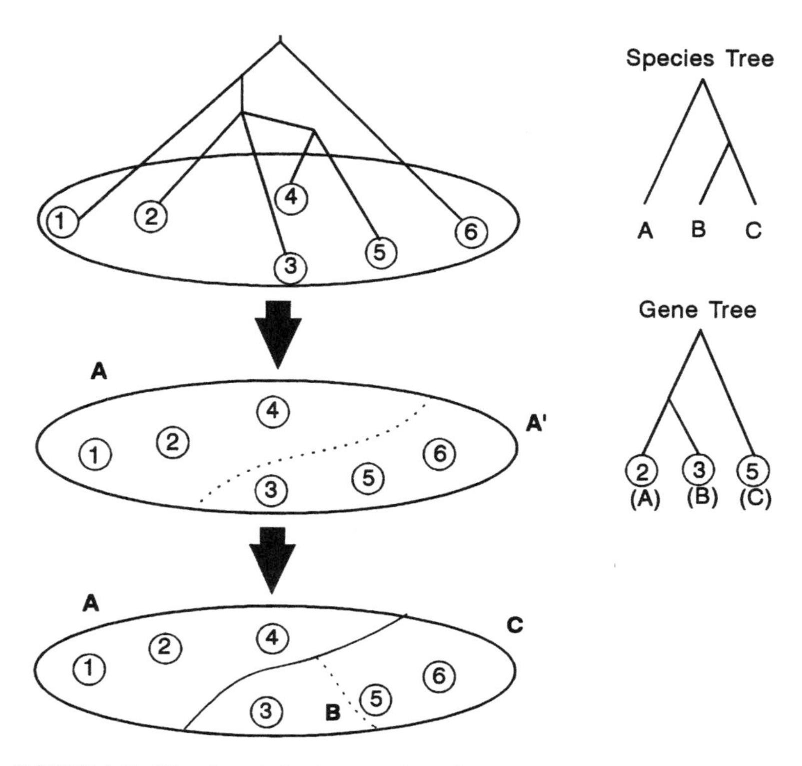

FIGURE 4-9. Vicarious inheritance of conflicting gene and species phylogenies. Descendant taxa inherit samples of alleles that are random with respect to the genetic lineage of those alleles. (Left) Geographic distribution of alleles (numbered 1–6) over ancestral range (ellipse), and vicarious inheritance of different allele combinations during vicariant form-making of species A–C (center and bottom). (Right top) Sequence of speciation for species A, B, C. (Right bottom) Speciation sequence implied by genetic relationships of alleles between species (from Ruvolo 1994: fig. 9; reprinted with permission of John Wiley & Sons).

netic relationships are not correlated with minimal geographic distance within vicarious distributions. Most challenging to phylogeny reconstruction are character affinities suggesting a close relationship between taxa geographically separated by other related vicariant taxa. These separated taxa occupy the peripheral "wings" of the distribution, and the pattern is called wing dispersal (the term "wing" refers to the lateral sides of a stage, and "dispersal" refers to vicarious differentiation and translation in space [Croizat 1994]). Peripheral character affinities may represent false phylogenetic relationships where plesiomorphic similarities obscure less apparent or undiscovered synapomorphies connecting each sister taxon to another adjacent taxon (Brundin 1981, Gray 1990),

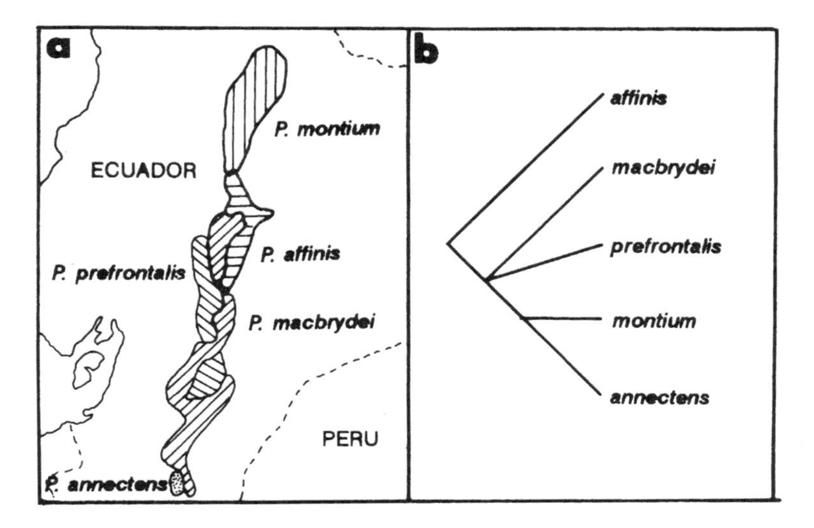

FIGURE 4-10. Wing dispersal. (a) Vicarious distribution of *Pholidobolus* lizards in Ecuador; (b) phylogenetic relationships. (From Hillis 1985: figs. 1, 5; reprinted with permission of *Systematic Biology*).

but cases of wing dispersal supported by characters that consistently qualify as being derived and uniquely shared are known.

Case Study 1: Ecuadorian Lizards

A phylogenetic analysis by Hillis (1985) illustrates a wing dispersal pattern for the two most derived species of *Pholidobolus* lizards in Ecuador (fig. 4-10). The genus comprises five vicariant species distributed along the eastern slopes of the Andes. The species are not separated by any physical barrier, and there is marginal sympatry between adjacent species distributions. The two most closely related species, *P. annectens* and *P. montium*, are located at the opposite ends of the generic range. This relationship is surprising because a continuous ancestral range would be required for the two lizards to share a most recent common ancestor. Hillis (1985) proposed that such a connection occurred when the other species were geographically more limited, and this direct connection was then lost following subsequent range expansion of the other species. The probability of such expansions was considered to be supported by recent observed distributional changes occurring in response to rapid changes in the current environment (Hillis 1985).

Consistent patterns of derived characters are often treated as having a one-to-one correspondence with speciation sequences, but if these character patterns also fail to correspond with geographic relationships, former migrations or range extensions such as those described for *Pholidobolus* are required to bridge the disjunction. For vicariant speciation to occur, it must also be asserted that the species are also in constant competitive equilibrium so that the expansion of one range is balanced by the contraction of another (Hillis 1985). This context is problematic because in many vicarious distributions, including these lizards, there is some geographic sympatry.

A vicariant form-making model suggests that the correspondence between the distribution of derived characters and the sequence of differentiation is not necessarily identical to speciation sequences within a particular ancestral range. As suggested by the recombination model, vicarious inheritance of molecular and morphological characters can be random with respect to the sequence of geographic differentiation (Ruvolo 1994). Wing dispersal patterns may be found in many different groups, but they are less frequent than patterns showing correlation between minimal geographic distance and phylogenetic relationship. It is possible that the main massings of ancestral characters contribute to a general correlation between geography and most parsimonious character sets, while incongruent characters represent smaller alternative massings. In this vicarious context wing dispersal is a special case where important ancestral main massings are incongruent with geographic sequences of differentiation.

Case Study 2: Eurasian Peonies

Wing dispersal patterns have been reported for cases of species hybridization in plants. Although hybrid species might be expected to occur in geographic proximity with their immediate parents, a study of the phylogeny and biogeography of *Paeonia* (Angiosperms: Paeoniaceae) reported the opposite situation (Sang et al. 1995). The genus comprises diploid and tetraploid species distributed from Western Europe to Japan, with most tetraploids occurring in the Mediterranean region (fig. 4-11). Phylogenetic relationships of the tetraploids were analyzed by sequencing nuclear ribosomal DNA, and evidence of hybridization was inferred from patterns of additivity for DNA sites between different diploid species.

Six different combinations of hybridization were recorded. One group comprising 10 western tetraploids involved multiple hybridizations of *P. japonica* or *P. obovata* with *P. lactiflora*, which are all restricted to eastern China and Japan. Disjunct parental diploids were also reported for three other tetraploid lineages, while only two polyploids were in relatively close proximity to their diploid progenitors. The current disjunct distributions between species of hybrid origin and their putative parents was described as intriguing. A plausible explanatory

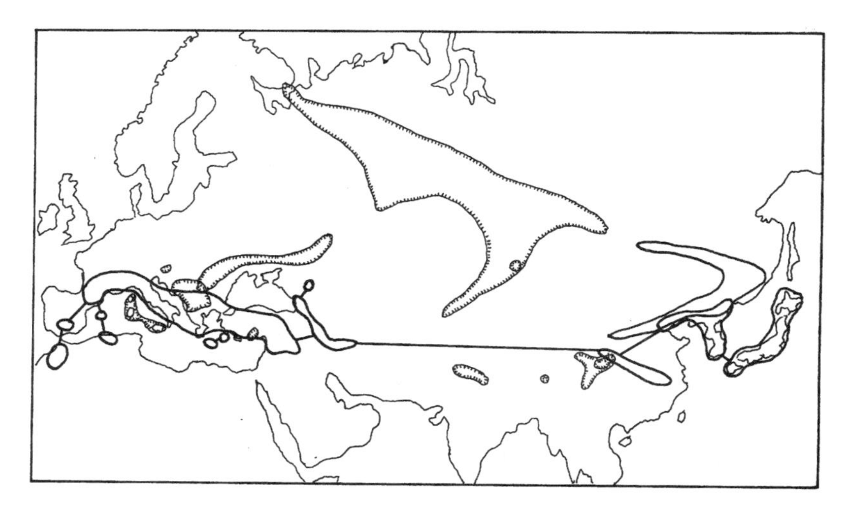

FIGURE 4-11. Wing dispersal of species of *Paeonia*. Track connects species distributions (thick lines) of 10 western tetraploids to proposed eastern Asian diploids parent lineages (after Sang et al. 1995: fig. 4).

scenario suggested former sympatry during times of Pleistocene refugia where the eastern Asiatic species must have had broader distribution ranges that included the Mediterranean. The Asiatic distributions were subsequently reduced during Pleistocene climatic changes, and their European ranges were replaced by their hybrids (Sang et al. 1995).

The biogeographic model of Sang et al. (1995) for *Paeonia* is similar to that of Hillis (1985) in proposing major levels of migration allowing for reorganization in distribution patterns. This model assumes that means of dispersal drive distribution, with disjunctions created by range contractions and then filled by migrations. To allow for redistribution while maintaining vicariant patterns, Hillis (1985) suggests competitive exclusion as a mechanism both for the formation of taxa as well as for their vicariism. Competitive exclusion is generally used to explain sympatric coexistence of ecologically similar species (Gotelli 1995). Its application to vicariism is problematic because many vicariants are not in direct contact, and in many situations vicariants also share some sympatry. The role of ecological processes such as competitive interactions between organisms and taxa is not excluded from vicariant form-making models, but if means of dispersal are not the driving force of geographic distribution, competition need not be invoked as a primary biological mechanism.

The lizards and peonies were each analyzed in isolation by the authors, and unique historical mechanisms were proposed to bridge each disjunction. Migration conflicts with the presence of vicariism, but a vicariant origin requires no radical alterations of distribution through migration. The application of this model to the relationships between genetic sequences and polyploidy such as that described by Sang et al. (1995) suggests that hybrids are not necessarily derived from extant taxa. The possibility of extinct species being involved was acknowledged for tetraploid taxa, and it is possible that the origins of the western tetraploids involves lineages that were also ancestral to the eastern Asiatic diploids. This would provide a direct geographic connection between eastern and western species, but through a common ancestor rather than the alternate contraction and expansion of the ranges of extant species.

4.4 Vicariant Form-making and Evolutionary Constraints

Vicariant patterns of spatial differentiation impose biogeographic constraints on the process of evolution. One implication of the reality of vicariant form-making to understanding evolution is that we must consider the possibility of processes that will allow an ancestor to change form over a wide geographic range and not at some specific point or center of origin. These processes will be more than an external influence such as natural selection because the evolutionary differentiation and diversification has proceeded in the face of many different environments. Characters and taxa are not randomly distributed, but comprise main massings of characters for individual taxa and spatial relationships shared by unrelated taxa in the form of standard or generalized tracks. These patterns suggest that biological significance of biogeographic constraints is not limited to the effects of local ecology on the fitness of individual populations. Biogeographic constraints imply a more general process of biological evolution involving mechanisms that are a consequence of biological processes that have been variously identified in evolutionary biology as morphological, biological, developmental, or phylogenetic constraints.

Laws of Growth

As geographic, morphogenetic, and phylogenetic fragments of a polymorphic ancestor, the evolutionary origin of individual vicariant taxa is also linked to the origin of related taxa. In this context local selection pressures do not act in isolation from constraints imposed by the spatial reorganization of ancestral distributions. Constraints represent a central concept in evolutionary biology and are usually invoked when biological patterns deviate from those expected by chance. Evolutionary constraints were recognized by Darwin (1860) as laws of growth that produce structures initially in no way advantageous to a species but that are subsequently taken advantage of by descendants (Craw 1984, Grehan 1984). Darwin (1888) believed that he neglected laws of growth because he was unduly influenced by the tacit assumption that every detail of structure, except for rudiments, was of some special though unrecognized service. Darwinian laws of growth are represented in modern biology by concepts such as phylogenetic, developmental, and genetic constraints (Maynard Smith and Vida 1990).

Developmental constraints on the phenotypic expression of genetic variation may limit the action of natural selection to variation that occurs along constrained pathways. Homologous genomic and developmental systems in related lineages may impose similar constraints on phenotypic variation. Through this process the same characters and phylogenetic trends can appear as parallelisms among vicariant clades (Simpson 1961, Grehan and Ainsworth 1985). Jacobs et al. (1995) recently applied this principle to the phylogenetic and developmental constraints on evolutionary patterns for coat color in tamarins (New World monkeys).

Phylogeography and Vicariant Evolution

Many molecular systematic studies of freshwater, marine, and terrestrial organisms, especially those conducted on mitochondrial (mt) DNA variability within species in the southeastern United States, have demonstrated that most taxa exhibit extensive polymorphism (e.g., Avise 1992). Avise et al. (1987) termed the comparative study of such DNA sequence variation across geography "phylogeography." The phylogenetic content in an mtDNA tree, when interpreted with observed geographic distribution of mtDNA clades, provides a picture of the phylogeographic structure of a taxon.

These studies have shown that major mtDNA phylogeographic patterns are shared across taxa. There is a remarkable degree of concordance in the major mtDNA phylogeographic disjunctions in species as diverse as the horseshoe crab, American oyster, toadfish, and diamondback terrapin. These concordant patterns may be interpreted as evidence of similar vicariant histories of population separation mediated by events in earth history (i.e., vicariant form-making in immobilism). In addition, shallower patterns of within-region phylogeographic structures due to recent dispersal and gene flow (i.e., mobilism) have also been observed (Avise 1994, Cunningham and Collins 1994).

Some molecular genetic processes involve evolutionary mechanisms that do not require an external selective force. For example, Kimura (1983) mapped the geographic distribution of alleles (at the alcohol dehydrogenase and esterase loci) in Japanese fishes. The biogeographic basis of the patterns is obvious from the maps provided by Kimura, who used them to argue for neutral (i.e., nonadaptive) vicariant evolution. Recent research on phylogeography and the concerted directional fixation of mutations indicates that the polytopic evolution of genotypes, rather than a genetic cause–effect relationship that is strictly ancestor–descendent, is one appropriate context in which to relate biogeography to molecular biology and evolutionary processes.

Concerted Evolution

"Concerted evolution" is the nonindependent evolution of repetitive DNA sequences resulting in a sequence similarity of repeating units that is greater within than between species (Dover 1982, Dover et al. 1993, Elder and Turner 1995). Each sequence repeat constitutes an independent unit that would, under the assumptions of neutral theory, be expected to diverge from the others, whether within the same species or in different species. The greater similarity of separate repeats within species suggests that organisms posses biological mechanisms that conserve sequence homogeneity within a family of repeated genes (Jinks-Robertson and Petes 1993).

Certain tandem arrays of repeated DNA are called satellite DNA. Accumulated mutations in satellite DNA among species appear too large to be constrained by selection, and intraspecific variation is too small to be accounted for by random drift. Whether or not satellite DNA has any functional role, it is significant for what it reveals about the molecular evolution of genomes. Nonfunctional satellite DNA exhibits the effects of molecular mechanisms in the absence of natural selection (Elder and Turner 1995), and the driving forces of molecular evolution often result from properties of the genetic material itself, rather than the effects of changes in the fitness of organisms (Crow 1987).

Patterns of concerted evolution are the result of a nonreciprocal transfer of sequence information between members of a gene family. Several genomic mechanisms are involved, the most prominent being unequal crossing-over and gene conversion. The continual stochastic or biased gain and loss of sequence variants can lead to the spread of a variant through a family and through a sexual population. The process of spread is called molecular drive, which can bring about, like natural selection and genetic drift, a long-term change in the genetic composition of a population (Dover et al. 1993).

The evolutionary population dynamics of molecular drive and concerted evolution remain largely unknown in natural populations. Most studies are descriptive, and only a few repetitive units have been studied, with little or no regard for processes such as reproductive isolation, genetic drift, or population status of the samples. Proposed affects of the satellite DNA can include the maintenance of proximity relationships of important genes on nonhomologous chromosomes, and in meiosis DNA repeats may be responsible for major genomic rearrangements ("macromutation") that may have selective or adaptive consequences, although any significant phenotypic impacts have yet to be confirmed (Elder and Turner 1995). Crow (1987) suggests that there is no inconsistency between molecular drive mechanisms and the neo-Darwinian principle that the evolution of form, function, and evolutionary adaptation are the result of natural selection. He acknowledged, however, that it is only after the intrinsic DNA processes have been operative that these new systems may be put to use. Dover (1993) suggests that the changing molecular basis of interactive molecules may be relevant to any type of interactions including cell adhesion, signal transduction, and recognition molecules involved with intercellular interactions that give rise to morphogenetic patterns.

The cohesive mode of change in concerted evolution and interactions with natural selection is an important consideration for the establishment of new biological functions (Dover and Strachan 1987). For aspects of the phenotype influenced by changes in multigene families, a slow phenotypic transformation of a population might permit the population to exploit previously inaccessible components of the existing environment. Functional consequences are affected by the way promoter genes (such as those sequences found in ribosomal (r) DNA which switch on or off the transcription of rRNA involved ribosome production) spread through a population. Promoter function is intimately linked to cellular activity and may be normally thought of as being constrained by strong selection against any change that negatively alters its function, but once it is amplified in the genome as multiple copies, there is the potential to relax selection (Ohno 1970). If one copy of the promoter is changed, the rest still provides sufficient levels of transcriptions. The new variant can become established without natural selection having any impact through a change in fitness, and molecular drive can continue to spread and homogenize the new variant through the population.

Molecular drive can, like selection, promote improbable long-term shifts in the mean genotype of a population, not as a result of a selective sorting of allelic variants resulting from their effects on individual fitness, but as a consequence of DNA turnover dynamics. Molecular drive mechanisms can operate on both single-copy genes and multigene families, and strict Mendelian genes and stable Mendelian populations in Hardy-Weinberg equilibria may not exist other than over short periods of time and in small numbers of progeny (Dover 1986). These molecular turnover mechanisms can have important functional consequences that arise from the spread of new promoters resulting in subsequent selection for molecular coevolutionary changes of key transcriptional proteins and their genes that functionally interact with these promoters. Because the extent of homogenization per individual is mostly

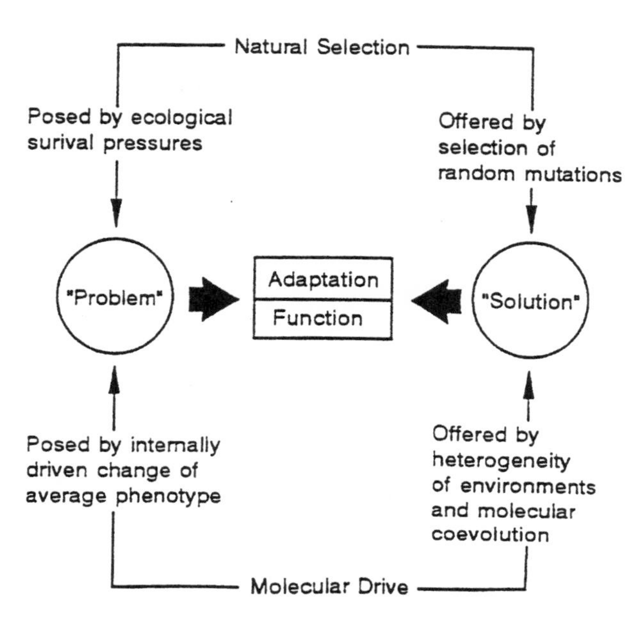

FIGURE 4-12. Conceptual model of organism–environment relationship with respect to natural selection and molecular drive in the origin of functions and adaptations. If biological functions are essential for the unique environmental relationships of a species, conventional theory would see the natural selection solution from random mutations to problems of organism survival. Molecular drive, in contrast, presents internally driven changes that are solved by the environment and an internal molecular coevolution of molecules maintaining interactions required of ontogeny (from Dover 1986: fig. 6; reprinted with permission of Academic Press).

uniform across a sexual population, the process would allow selection to favor alleles that are better able to interact with the gradual influx of new promoters. Through an internal coevolution of the molecules, the continual maintenance of developmental interactions is ensured (Dover 1986, 1992, 1993).

Models of adaptation based solely on natural selection treat selection as the mechanism for solving problems of survival imposed by the environment, but with molecular mechanisms the sequence could be reversed. If molecular-driven changes are tolerated by a preexisting environment, the environment may be treated as a solution to the problem set by the transformed population (fig. 4-12). The final observed function would be due to the population adopting a new environmental component. In this sense, the adaptation is non-Darwinian. Intermediate stages of this adaptation may involve a succession of appropriate environments that are open to exploitation at each stage (Dover 1986).

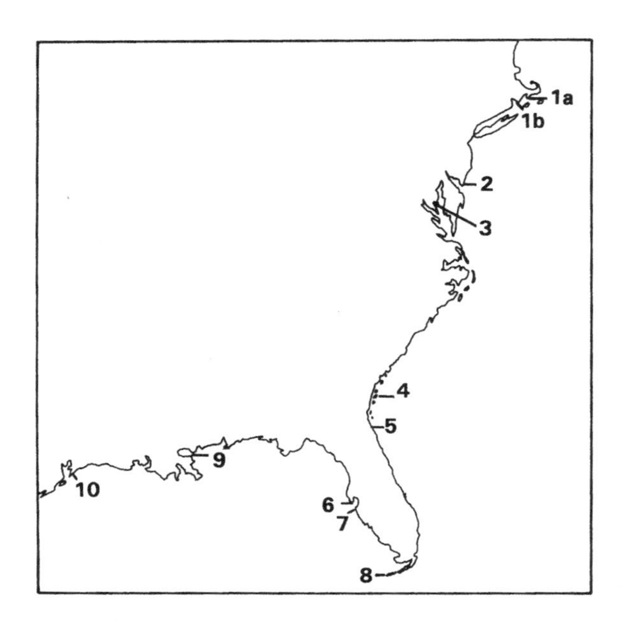

Population sample No.

	2	3	4	5	6	7	8	9	10
1	35.0	48.9	71.6	71.1	69.4	66.5	41.1	70.3	74.2
2		42.4	41.8	41.4	36.3	38.5	50.2	36.5	44.6
3			52.5	52.3	53.5	48.0	47.6	54.8	50.2
4				88.0	92.3	77.5	47.7	94.2	80.5
5					90.3	80.1	47.6	90.5	81.3
6						83.2	48.2	95.9	88.5
7							35.8	80.1	73.4
8								43.9	37.4
9									84.4

FIGURE 4-13. Concerted evolution of *Hind*III satellite DNA sequences in populations of pupfish (*Cyprinodon variegatus*). (Top) Geographic distribution of samples; (bottom) pairwise proportional similarities of *Hind*III canonical satellite monomers among population samples (after Elder and Turner 1994: fig. 1; reprinted with permission of the National Academy of Sciences).

Case Study: Atlantic Coast Pupfish

For concerted evolution to have a significant evolutionary impact on differentiation, intraspecies differences in repeat patterns of DNA are expected. In a geographic study of a pupfish (*Cyprinodon variegatus*) distribution along the Atlantic coast (Elder and Turner 1994), local samples were found to show significant differences in DNA repeat patterns (fig. 4-13). Repetitive cloning and direct genomic sequencing experiments failed to detect significant intrapopulation or intraindividual variation, suggesting high levels of sequence homogeneity within populations for *Hind*III DNA sequences.

Patterns of within-species concerted evolution such as that represented in the Atlantic pupfish imply a state of immobilism where movement of individuals is localized and spatial isolation is not dependent on absolute geographic distance. Pupfish levels of sequence divergence were not correlated with minimal geographic distance, and, although the two most similar samples (95.9% similarity) were geographically adjacent, the two most divergent samples (35% similarity) were also separated by a similar distance. This geographic pattern is compatible with complex geographic patterns of vicariant differentiation within an ancestral range, particularly if this differentiation involves periods of limited or local mobilism between adjacent populations within the distribution range. The geographic patterns reported by Elder and Turner (1994) may suggest that concerted evolution between species may be initiated as a process of vicariant differentiation within the ancestral distribution range. If absolute geographic distance is not always correlated with patterns of concerted evolution, phylogenetic affinities may range from those that are directly correlated with nearest allopatric neighbors (vicariance criterion) to those that are incongruent (such as wing dispersal).

4.5 Conclusions

Although geography may come into consideration in systematics, it is evident that an accurate biogeographic analysis depends on the best systematic hypothesis available. Under this requirement, biogeography has often been regarded as a derivative science lacking independence from the prerequisites of phylogeny and taxonomy. The flow of information becomes unidirectional, from phylogenetic systematics, which is the basis for biogeographic analysis, to biogeography, which becomes merely a recipient (Ferris 1980, Nelson and Platnick 1981, Patterson 1981, Humphries 1990, 1992). Taxonomy alone is qualified to construct accurate classifications that are "independent of distribution in time and space" (Platnick and Nelson 1989).

Every identifiable evolutionary unit, whether population, species, or higher taxon, must be uniquely related to spatiotemporal particulars. Our experience of these life cycles, lineages, or units is not of their existence solely in time, but of them as spatiotemporal entities (Ghiselin 1974, Hull 1976). If taxa are spatiotemporally bounded entities, then biogeography must be a necessary consideration in evolutionary, phylogenetic, and systematic studies, and a natural classification can only be established after, not before, a biogeographic analysis. Analysis of the spatiotemporal coordinates of taxa is needed in systematics for the relation between space, time, and form to aid in elucidating phylogenetic relationships and assist in reconstructing phylogeny (Croizat 1958, 1964; Fisher 1994).

Geographic patterns should reveal something about the way that life has evolved, as phylogeny is not independent of space and time. From a philosophical basis, sources such as the geographical distribution of taxa cannot be excluded as irrelevant in phylogeny reconstruction (Hull 1988, Sober 1988). Panbiogeography recognizes a phylogenetic role for geography within which the dependence of biogeography upon systematics is acknowledged, but so too is the reliance of systematics on geographic information. Panbiogeography does not represent a claim to a necessarily superior insight into the evolution of taxa, but it does suggest that an explicit consideration of biogeography, past and present, is essential to inquiries into the history of life.

Biogeographic track analyses suggest that the geographical context is an important component of phylogenetic evolution. Vicariant form-making is spatial and involves the recombination and redistribution of characters through vicarious differentiation of widespread polymorphic ancestral populations. Evolutionary differentiation involves biology, history, and geography where the common center of origin lies beyond any single ancestral taxon, area of endemism, or character state, within multiple points in series of biogeographic and molecular tracks and nodes. There is no longer a simple origin.

5

Tracking the Trees of Life

Line, Map, and Matrix

I want chiefly a monster-map (on the conical projection)—so as to map out the data and put them to the test of discussion.

S. P. Woodward (letter to C. Darwin, 4 June 1856)

Could you colour a map of the world with a few primary tints & shade off: or could you give a graphical representation by concentric lines crossing where floras intermingle.

C. Darwin (letter to J. D. Hooker, 7 March 1855)

Descriptions that we make of the biogeographical world (e.g., geographic distribution maps of taxa, floral and faunal lists) do not speak for themselves; biogeographers must give them meaning through analysis and interpretation. Since the late 1960s the quantification of methodology (e.g., Sneath 1967) and the development of more accurate forms of historical representation have become key goals in the search for rigorous methods to analyze phylogenetic relationships and distributional data (Ball 1976). This increasing use of quantifiable cladistic and phenetic classificatory methods and statistical hypothesis testing in biogeography and systematics represents a tremendous step forward. Their applications have forced biogeographers toward a more precise formulation of methodological practices and theoretical ideas and the exact quantification of their implications.

A sound basis for understanding geographical distribution patterns and the processes governing them has begun to emerge through a series of successive tentative approximations and ongoing technical debates (Morrone and Crisci 1995). Biogeography requires this constant dynamic of theoretical debate, quantitative development, and new techniques of graphical representation that identify multiple directions to move in, as well as rigorous observations of life's distribution that can help constrain our choice of those directions.

5.1 Representations of Geographic Space in Biogeography

Observed geographic distribution patterns involve "movement" or "translation in space" and a spatial dimension represented by locality records, so geographic methodology as well as the pure biology of biogeographical phenomena must be considered (Morain 1984). Spatial data represent observations on geographic individuals and objects that can only be interpreted satisfactorily when their locations are taken into consideration. Insights into spatial processes may be gained from a study of spatial form. The distribution of collection and recorded localities in space, their relative position with respect to each other, and the links that appear to exist between them are one basis for biogeographical analysis.

Cartographic Representation in Biogeography

Maps are the most familiar and general representation for geographic phenomena. The initial database for biogeographical analysis is a compilation of mapped collections and/or recorded localities and distributional areas for the taxa under study. Comprehensive discussion of areographic (the depiction of distribution areas) and cartographic representation in biogeography can be found in Udvardy (1969), Stott (1981), and Rapoport (1982).

In most biogeographical studies and taxonomic monographs, one of two basic mapping conventions is adopted: dot maps, in which every collection and/or recorded locality for a taxon is mapped, or outline maps, in which the collections and records are summarized and the whole range of a taxon is enclosed within a line to represent a distribution area. The type of mapping technique used depends on the primary purpose of the maps.

Dot maps have the advantage of summarizing a great deal of information, but it is impossible to illustrate many taxa clearly on one map. This is a major problem if the aim of the study is to compare distributions and summarize the main features. Dot maps can include more information than is needed for comparative studies, especially when the geographical scale is regional or global. Outline maps have the advantage that the distributions of many different taxa can be illustrated on one map. A problem with these maps for analytical and comparative purposes is that the outlines can become so generalized that important localities are not sufficiently highlighted (Heads 1994a).

A compromise approach is to represent the distribution as an outline, but to draw the outline through actual localities that are represented as dots. In a situation where a taxon has a broad distributional range, the outline can be drawn through the outlying localities. This approach allows putatively or known related taxa to be shown on the same map, avoiding the problems of map treatment of taxa in alphabetical order, which are difficult to use for comparative purposes (Heads 1994a). Once mapped, taxa are described as vicariant or vicarious if they are spatially isolated or display minimal geographic overlap (parapatry) (Cain 1944, Udvardy 1969).

5.2 Graphical Representation in Biogeography

Representation in biology aspires to the condition of a graph (G. Myers 1990). The application of graphical approaches to biogeographical studies has great potential because images such as graphs are fundamental in reducing the complexity of representation, analysis and interpretation in many fields. Biogeographic track graphs are one way of reducing the complexity of distribution maps into a form where it might be possible to capture analytically any pattern or structure in the initial data.

A track is a graphical representation of the spatial form of a given biogeographic distribution. It is a line graph drawn on a map of the geographic distribution of a particular taxon (e.g., species, species group, genus, family, etc.) that connects the collection and/or recorded localities and distribution areas of that taxon or the subordinate taxa belonging to it (fig. 5-1). This track is interpreted as a graph of the geographic distribution of that taxon and represents the "primary coordinates" of that taxon in space (Croizat 1964). Construction of these track graphs permit a precise comparison between tracks for individual taxa and forms the basis for detailed analyses of distribution patterns.

Croizat (1952, 1958, 1964) is usually credited with originating the use of track graphs in biogeographic analysis, but attempts to represent geographical distribution as line graphs or tracks are known that pre-date his work. Cernosvitov (1936), in a study of earthworm distribution, represented distributions as vertices and links between those vertices as lines or edges (fig. 5-2a). Skottsberg (1936) linked the distribution localities of phylogenetic lineages in the plant genera *Astelia* and *Collospermum* by line graphs (fig. 5-2b). These trends toward graphical abstraction were carried a stage farther by Townsend (1942), who reduced geographical distribution areas to vertices and represented paths of biotic spread as links or graph edges (fig. 5-2c).

Recent applications of the track method on a variety of spatial scales (fig. 5-3) include analyses of the trematode parasites of frogs (Anura) (Brooks 1977), Caribbean and Central and South American flies (Matile 1982), freshwater molluscs (Taylor 1988), upper Cretaceous and Palaeogene angiosperms (Taylor 1990), freshwater cave crustacea (Holsinger 1989), and freshwater river and stream crustacea (Morrone and Lopretto 1994). Use of the biogeographic track mode of data representation is

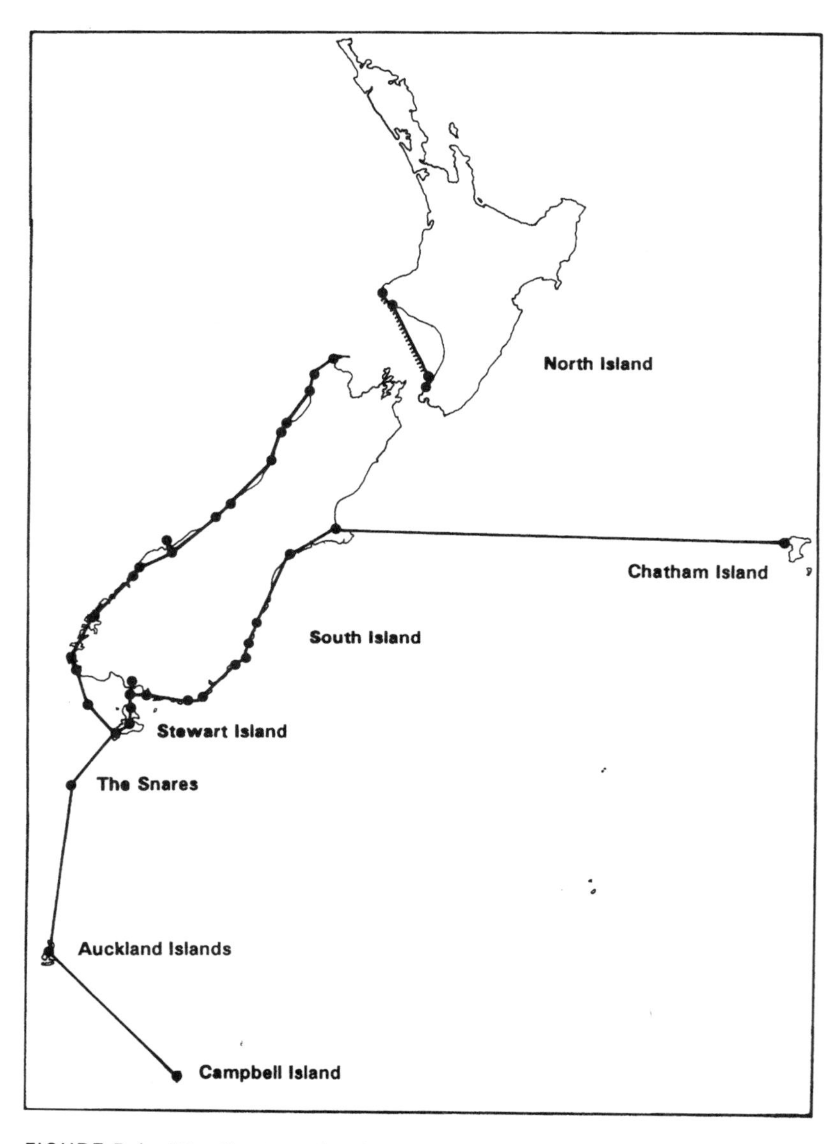

FIGURE 5-1. Distribution of *Hebe elliptica* in New Zealand represented as a track graph formed by connecting the known localities by a line graph (solid line, *H. elliptica* var. *elliptica;* hatched line, *H. elliptica* var. *crassifolia*) (after Heads 1993: fig. 11).

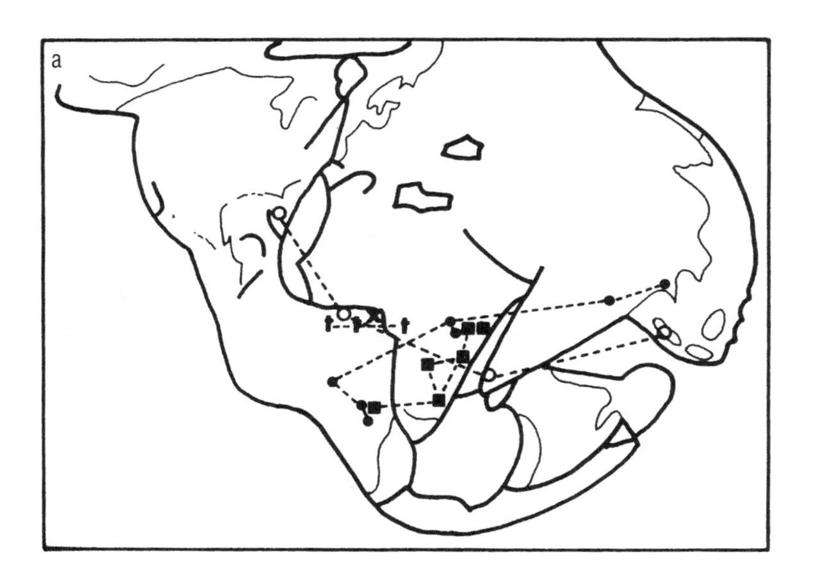

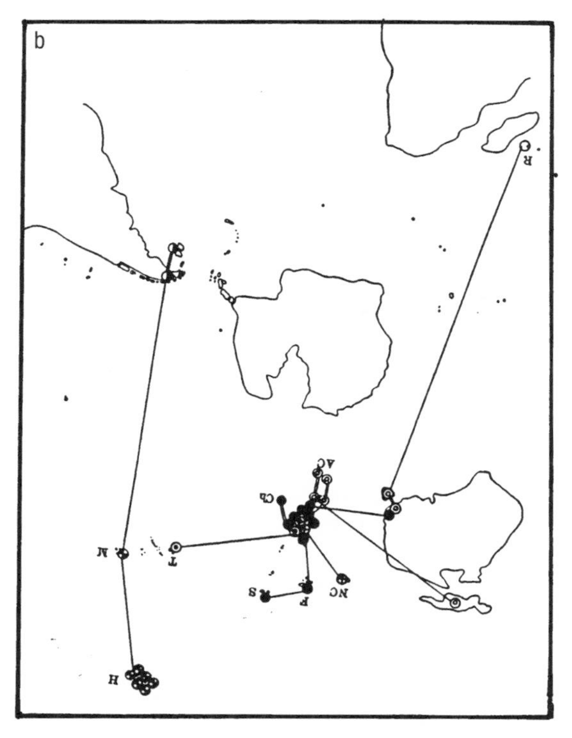

FIGURE 5-2. Representation of geographic distributions as line graphs or tracks that pre-date panbiogeography. (a) Distributions as graph vertices and tracks as graph edges for earthworm distribution (after Cernosvitov 1936: fig. 1); (b) line graphs representing the distributions of sections of *Astelia* and *Collospermum* (Angiosperms) (from Skottsberg 1936: fig. 1; re-pinted with permission of the University of California Press); (c) geographical areas as graph vertices and paths of biotal spread from the Antarctic as graph edges (after Townsend 1942: plate 2).

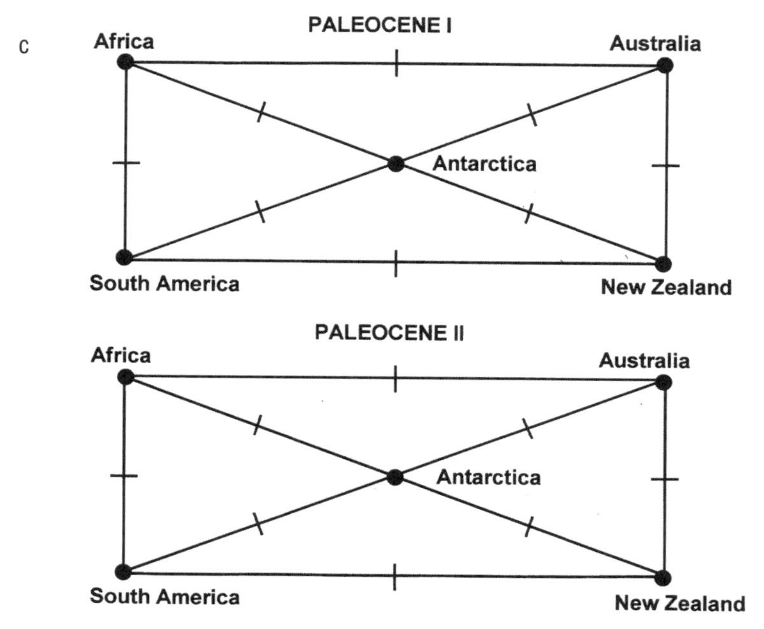

FIGURE 5-2. (*continued*)

widespread, and it has been applied to a diverse range of terrestrial, freshwater, and marine organisms (table 5-1). Recently, Fortino and Morrone (1997) have proposed standardized signs for the representation of track analyses based on graphic design principles.

Graphical Representation of Vicariance Zones

A vicariant distribution (the occurrence of two or more closely related populations or taxa that occupy mutually exclusive distribution areas) can be condensed into an integrated graphical representation—a literal diagram, as a line between two distribution points. This is the sense in which the term "track" was used by Nelson (1973:312):

> The nature of a track may be illustrated with reference to the two species of the subfamily Heterotinae (Teleostei, Osteoglossidae): *Arapaima gigas* of the Amazon basin of South America, and *Heterotis niloticus* of tropical west Africa and the upper Nile. If one can decide that the track is to extend from one specific point to another, let us say from the mouth of the Amazon to the mouth of the Niger, one may connect these points by the shortest possible line—a track extending directly across the tropical Atlantic Ocean, more or less in parallel with the equator.

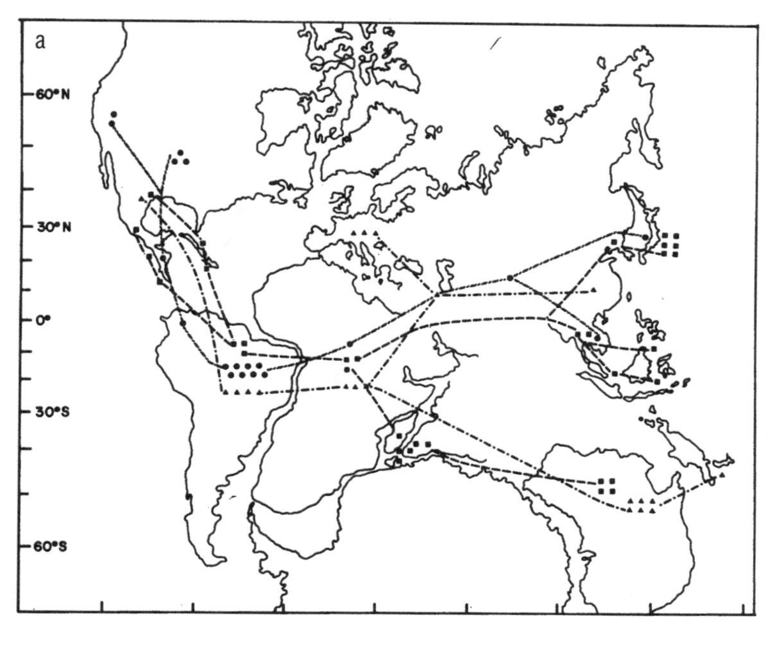

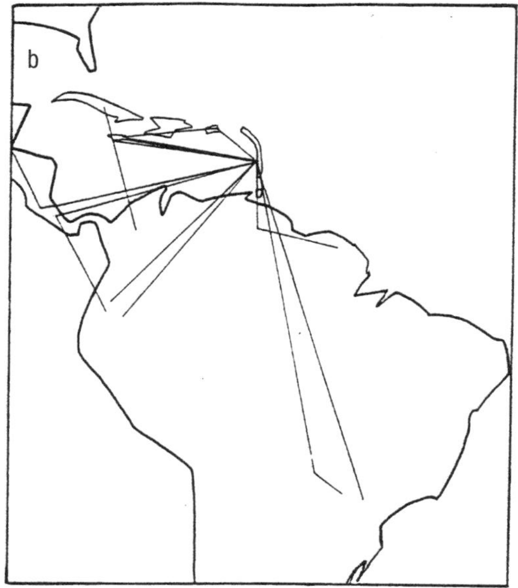

FIGURE 5-3. Examples of recent applications of the track method. (a) Trematode parasites of Anura (from Brooks 1977: fig. 3; reprinted with permission of *Systematic Biology*); (b) South and Central American and Caribbean flies (from Matile 1982: map 2; reprinted with permission of *Bulletin du Muséum d'Histoire Naturelle*); (c) bipolar freshwater molluscs (from Taylor 1988: fig. 6; reprinted with permission of Elsevier Science); (d) freshwater cave crustacea (from Holsinger 1989: fig. 5; reprinted with permission of author); and (e) freshwater crustacea (after Morrone and Lopretto 1994: fig. 19; reprinted with permission of *Journal of Biogeography*).

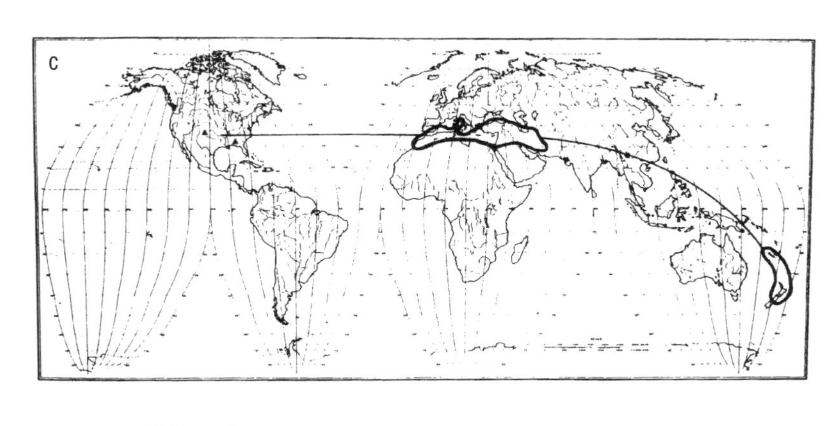

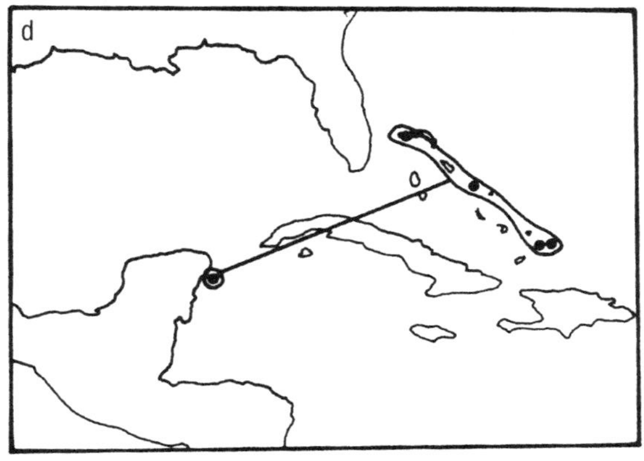

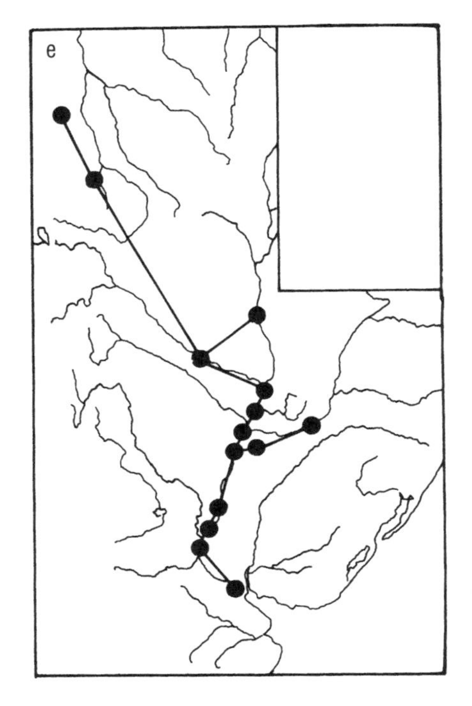

FIGURE 5-3. (*continued*)

TABLE 5-1. Some examples of the use of track modes of analysis and/or representation.

Geographic area	Taxonomic/ community group	Reference
Caribbean	Freshwater Crustacea	Holsinger 1989, 1991
Caribbean	General	Rosen 1976, 1978
Central America	Freshwater crabs	Rodriguez 1986, 1992
East-Pacific, Central America	Marine Crustacea	Myers 1988, 1994
East Wallacea	General fauna	Michaux 1994
Falkland Islands	Plants and insects	Morrone 1992
Indo-Pacific	Perches	Vari 1978
New Zealand	Calanoid copepods	Jamieson 1998
New Zealand	Grass moths	Patrick 1991
New Zealand, world	General	Matthews 1990
North America	Caddisflies	Hamilton and Morse 1990
North America, Europe	Pselaphid beetles	Carlton 1990, Carlton and Cox 1990
North America, world	Fresh water Mollusca	Taylor 1960, 1988
North and Central America	Cave collembola	Christiansen and Culver 1987
Northern Hemisphere	Garfish	Wiley 1976
Pacific rim	General	Sluys 1994
Pantropical	Araceae angiosperms	Mayo 1993
South America	Leptodactyloid frogs	Lynch 1986
South America	Freshwater decapods	Morrone and Lopretto 1994
South and Central America	Flies	Matile 1982
Southeast Asia	Hesperid butterflies	Chiba 1988
Southern Hemisphere	Proteaceae angiosperms	Weston and Crisp 1996
Southern Hemisphere	Southern beeches and parasites	Humphries et al. 1986
Southern United States	Freshwater fish	Mayden 1985
South Pacific, New Zealand	Thymelaeaceae angiosperms	Heads 1994d
South Pacific, New Zealand	Scrophulariaceae angiosperms	Heads 1993, 1994a,b,c
Southwest Pacific	General	Craw 1979
Southwest Pacific	Micropterygid moths	Gibbs 1983, 1990
Southwest Pacific, world	Caddisflies	Henderson 1985
Southern Hemisphere	Vertebrates	Cracraft 1980
Southern S. America/ Falklands	Rhytirrhinini weevils	Morrone 1993
World	Frogs and trematodes	Brooks 1977
World	Crocodiles and helminths	Brooks 1979
World	Moths and butterflies	Grehan 1991a
World	Gadoid fish	Howes 1990
World	Marine bugs	Anderson 1982
World	Grasses, reeds, Poales/ Restionales	Linder 1987
World	Fossil angiosperms	Taylor 1990
World	Relict weevil taxa	Morrone 1996

Kavanaugh (1980) used this notion of a track to illustrate vicariant relationships among taxa in the carabid genus *Nebria* (fig. 5-4a). Dots representing the most proximate localities were connected by solid lines that traverse the intervening gaps between disjunct geographic distributions of related taxa. These tracks are equivalent to what Perkins (1980) termed "vicariance zones" in his study of New World hydraenid beetles (see also Noonan 1989, Spanton 1994). Anderson (1982) and Gibbs (1990) have used this method to identify congruent patterns of vicariance zones in studies of insect biogeography (fig. 5-4b,c), and Howes (1990) applied the method in gadoid fish biogeography (fig. 5-4d).

Generalized Tracks

A generalized (or standard) track is a set of congruent, overlapping individual tracks. It is a line graph summarizing the replicated distribution patterns of different groups of organisms. This generalized track is an empirical description and summary of a variety of individual tracks.

Consider, for example, the distributions and individual tracks of a flightless anthribid species pair (fig. 5-5a), a flightless litter weevil species group (fig. 5-5b), and a wind-dispersed tree-daisy group (fig. 5-5c). By superposing the individual tracks for each of these three taxa, they are seen to be congruent tracks to the extent that the track pattern for each group either replicates or nests within the tracks of the others (fig. 5-5d). The line graph summarizing all the individual graphs is called a "generalized" or "standard" track.

It must be emphasized that no assumptions concerning the biogeographic processes responsible for the congruent tracks are implied. They could represent (1) the track of an ancestral biota that was subdivided by vicariant events, (2) a concordant dispersal pathway that was used concurrently by all three taxa, (3) three separate dispersal events, or (4) a combination of these scenarios.

5.3 Quantitative Track Analysis

Having established that a literal graphical approach might prove useful in biogeography, it is now necessary to demonstrate that track methods can have a strictly comparative, quantitative, and statistical basis. In addition, statistical hypothesis testing is necessary to bring the discipline of historical biogeography up to the scientific standards that apply in ecological biogeography and to determine how much similarity between tracks for individual taxa is significant. These considerations require a quantifiable basis for track analysis and explicit statements of

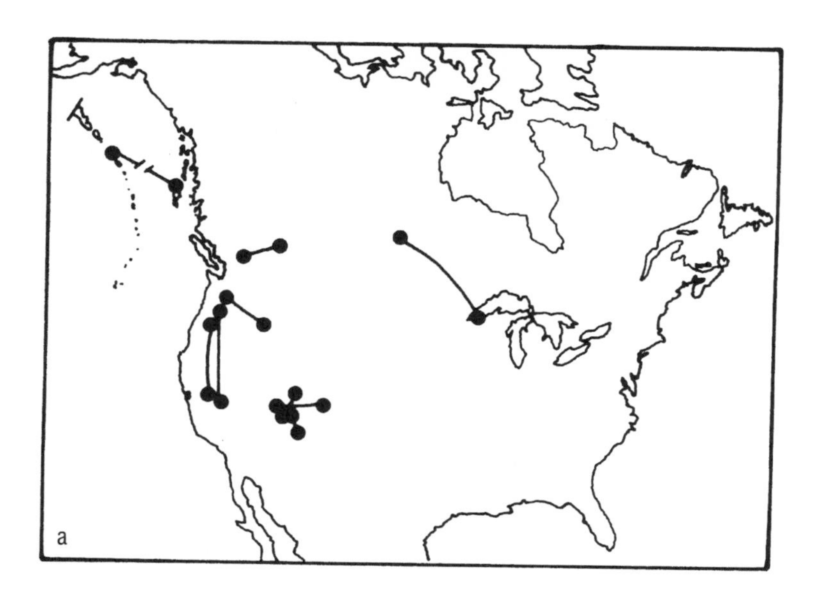

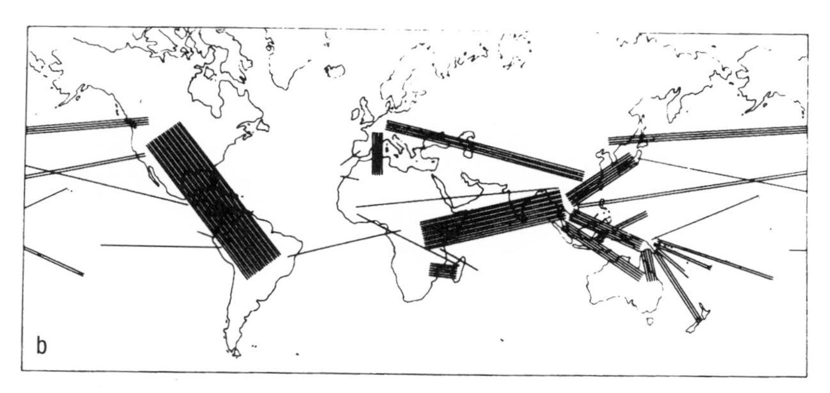

FIGURE 5-4. Track representation of a vicariant relationship as a line between two distribution points. (a) Carabid beetles. Dots representing the most proximate localities are connected by solid lines for sister taxa (from Kavanaugh 1980: fig. 15; reprinted with permission of *The Canadian Entomologist*); (b) semiaquatic bugs (Gerromorpha) (after Anderson 1982: fig. 63); (c) moths (from Gibbs 1990: fig. 7a; reprinted with permission of SIR Publishing); (d) gadoid fish (from Howes 1990: fig. 15; reprinted with permission of *Journal of Biogeography*).

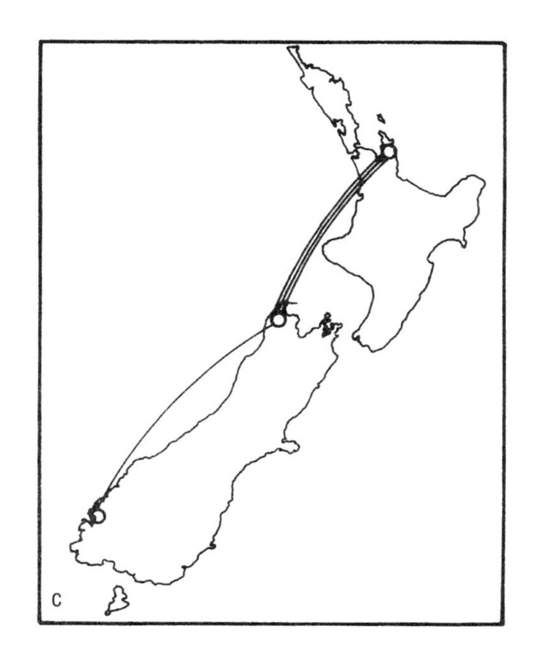

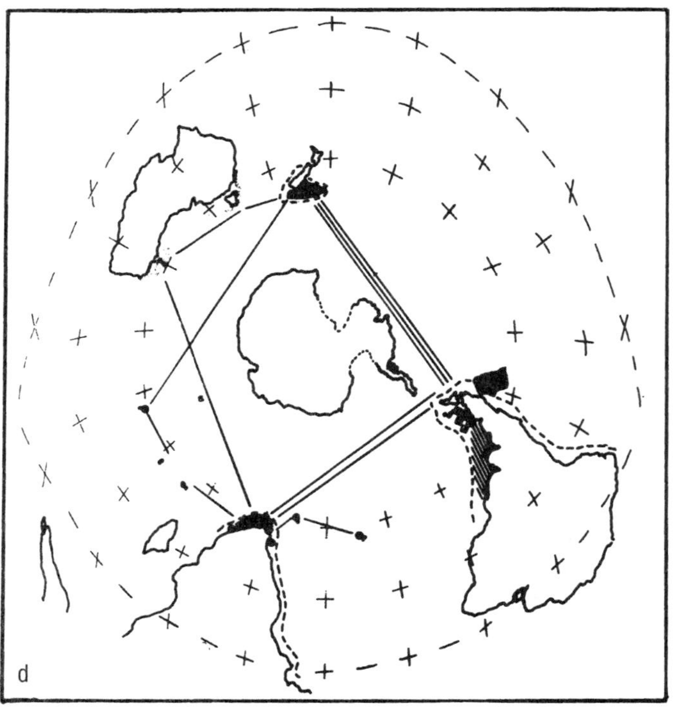

FIGURE 5-4. *(continued)*

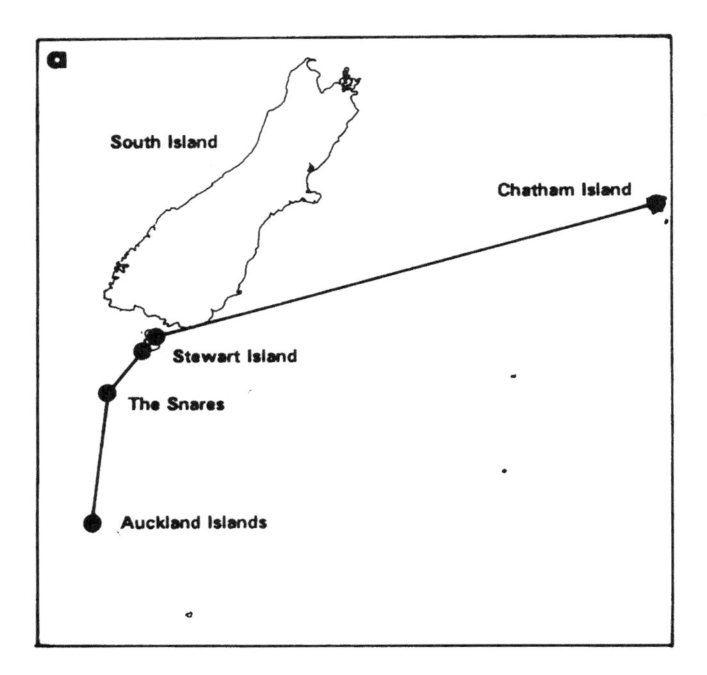

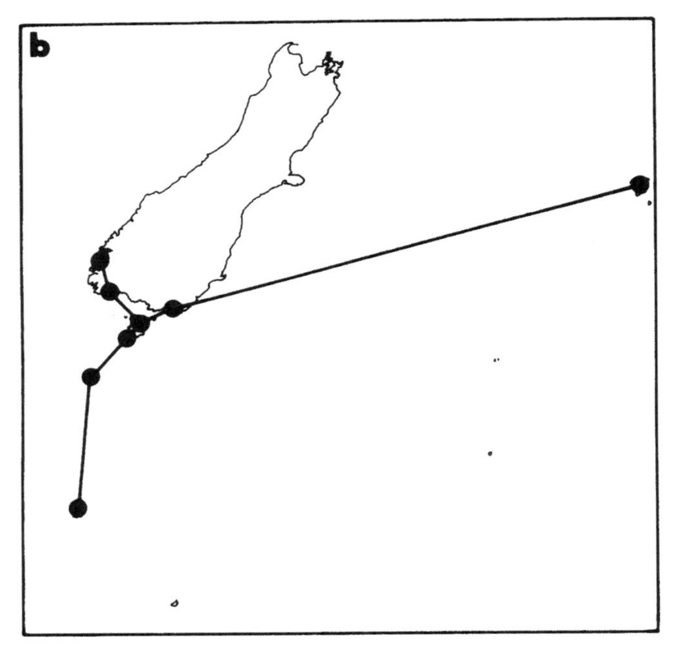

FIGURE 5-5. Construction of a generalized or standard track graph for taxa distributed in southern New Zealand and the Chatham and subantarctic islands. (a) Individual track of related anthribid species (*Caecephatus huttoni* and *C. propinquus*); (b) individual track for related species of flightless litter weevils (*Exeiratus setarius* species group); (c) individual track for related species of tree daisy (*Olearia* species group A); (d) the generalized or standard track summarizing individual tracks a–c (from Craw 1989: fig. 11; reprinted with permission of *Systematic Biology*).

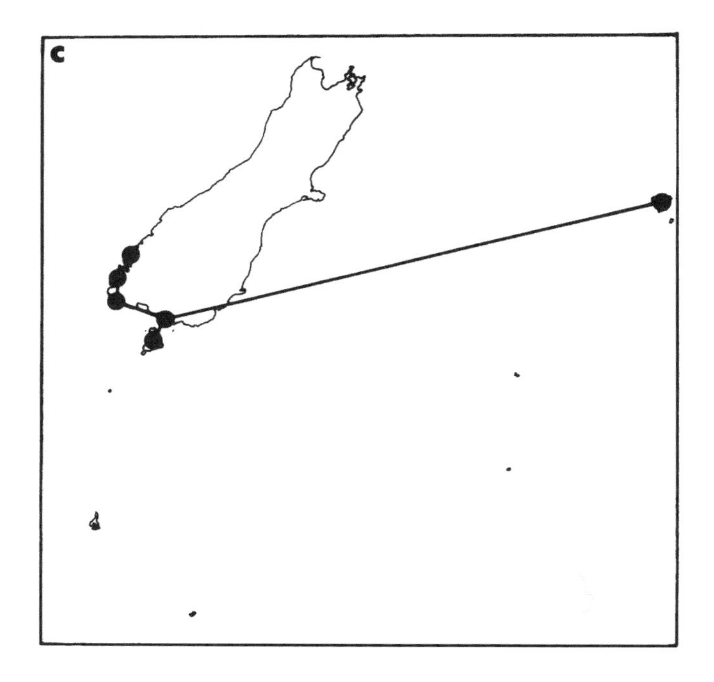

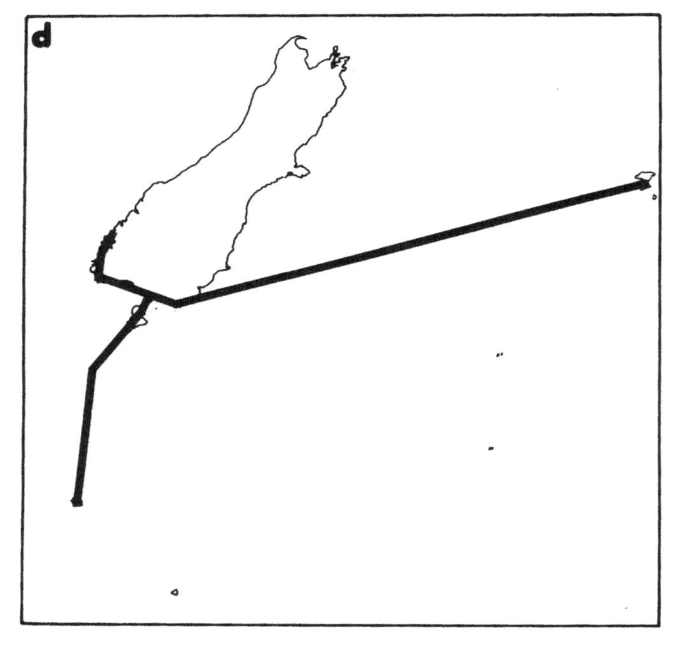

FIGURE 5-5. *(continued)*

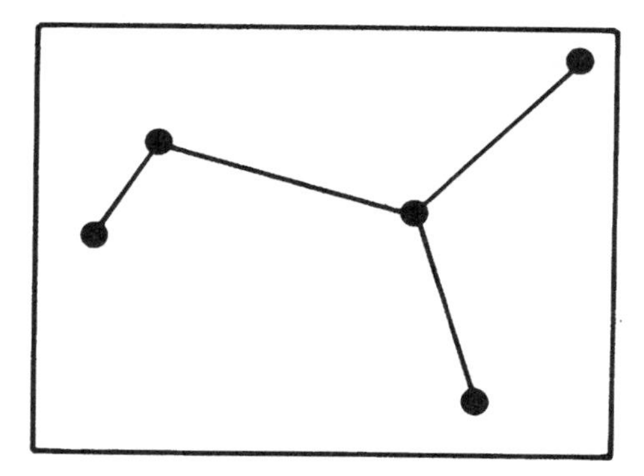

FIGURE 5-6. Representation of a track as a spanning tree for five points and four track links. When the points are connected in a sequence giving the minimum total length possible, the tree corresponds to a minimal spanning tree (from Page 1987: fig. 3b; reprinted with permission of *Systematic Biology*).

null hypotheses with criteria for rejection that can be directly incorporated into empirical track studies.

Conventional statistical tests are inappropriate in historical biogeographic and evolutionary studies because the available data do not meet the assumptions of those tests (Pagel and Harvey 1988, 1991). Instead, randomization tests are used to represent the possibility of chance individual dispersal for each taxon as a null hypothesis (Manly 1991). Attempts to develop a quantitative, repeatable, and statistical basis for track analysis were pioneered by McAllister et al. (1986), Page (1987), Connor (1988), Craw (1989), and Henderson (1990, 1991). These analytical developments used principles and practices derived for the most part from graph theory to provide objective and quantitative methods for drawing and comparing tracks. In addition, they provide a means for proposing statistical tests for hypotheses of track congruence and correlation.

Minimum Spanning Tree Method

One procedure for quantifying tracks is based on the concept of a minimal spanning tree. A minimal spanning tree (MST) is an acyclic graph that connects all localities or distributional areas of a taxon in such a way that the sum of the lengths of the links connecting each locality or distribution area is the minimum possible. For n localities or distribution areas, the line graph connecting them will consist of n–1 links (fig. 5-6). If the taxon under consideration is differentiated into distinct taxa (e.g., vicariant species in a genus), an MST can be constructed for each distinct taxon and then these separate graphs may be linked to form a single MST. In many situations these links may not represent accurate estimates of the true distance between those localities, particularly when taxa are distributed over a large area, because of the curvature of the earth's surface. The formula of Phipps (1975) can be used to correct for this distortion.

The construction of a track graph by connecting nearest geographic neighbors to form an MST is a methodological rule only in the absence of other information such as evidence of phylogenetic relationship. Procedures for constructing MSTs for taxa where either partially or fully resolved phylogenetic trees exists have been proposed (Page 1987, Craw and Page 1988, Henderson 1990), and a statistical method for testing the robustness of MST analyses has been developed (Henderson 1990). Advocacy of the MST procedure is not a claim that spatial evolution is parsimonious (Page 1987). This method requires no methodological assumptions of hierarchical relationships between localities or distribution areas and directly incorporates spatial information as part of the analysis (Craw 1983).

One reason for choosing MSTs over other types of graphs as a preliminary procedure for track construction is that an MST can be calculated exactly and efficiently, on an objective and repeatable basis by any number of people who will generate the same MST from the same data (Rapoport 1982). Spanning trees that do not have a minimal length result in many different possible combinations, but without any criteria for choosing one tree over any others.

The MST procedure can be summarized by the following steps:

1. Plot the locality records of each taxon on a map (fig. 5-7a),
2. Connect all the localities of each taxon by a minimal spanning tree (fig. 5-7b),
3. Connect each of the minimal spanning trees for all taxa by further minimal spanning links (fig. 5-7c).

Through this approach to track analysis, the cartographic display of track relationships can become more rigorous and standardized. Minimal spanning trees are useful exploratory devices that give a good visual indication of distributional data (Bailey and Gatrell 1995). An MST combines the detailed information content of plotted locality records with an easily decoded broad picture of their distribution—their limits, disjunctions, outliers, and cores of continuous populations. These features of taxon distribution, especially in complex situations involving large numbers of localities on a broad geographic scale, may not be so easily recognized using more traditional methods with unconnected dots for each locality or outlines enclosing the known range (Henderson 1990).

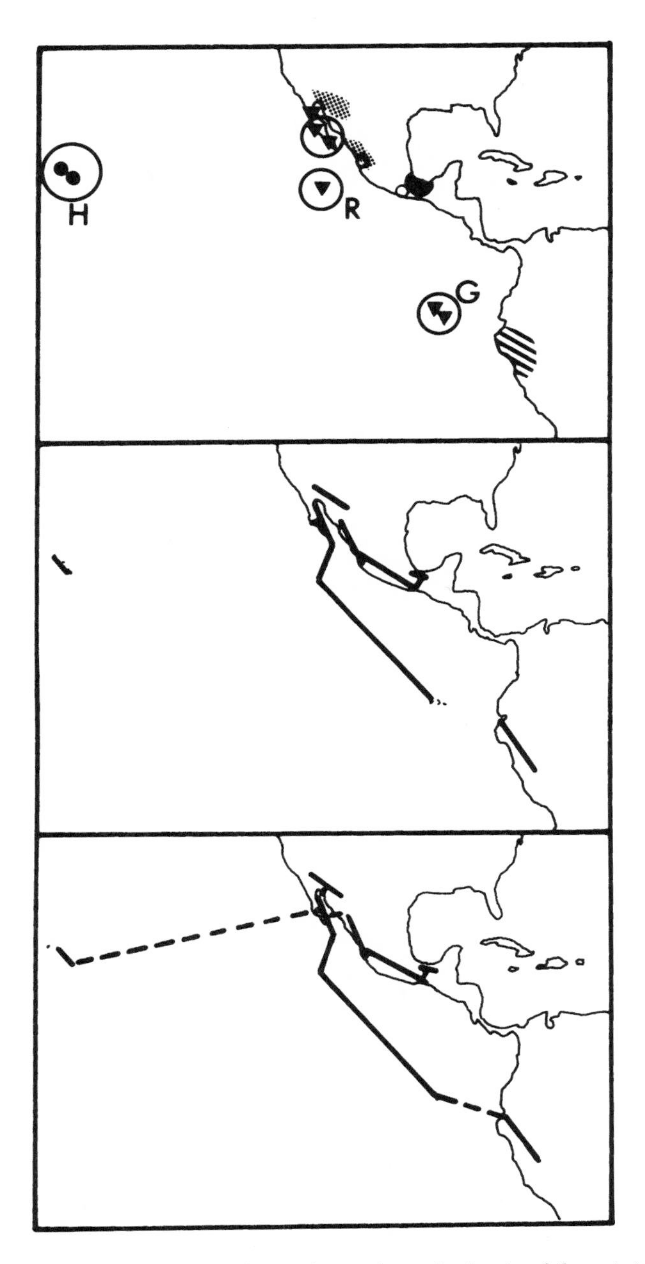

FIGURE 5-7. Construction of a track graph on the basis of the minimal spanning tree (MST) criterion for a hypothetical group of related taxa distributed in Hawaii (H), the Revillagigedo Islands (R), southwestern North America, Central America, the Galapagos Islands (G) and northwestern South America. (top) Plot the distribution of each taxon, (middle) draw a MST for each taxon, (bottom) connect each taxon to the other taxa by a MST (from Page 1987: fig. 2; reprinted with permission of *Systematic Biology*).

Parsimony and Minimal Spanning Trees

Parsimony is a methodological convention for choosing from a set of possible statements. In cladistics, parsimony is a rule for choosing a subset of trees from the total set of $(2n-5)!!$ possible Steiner trees (tree graphs connecting a set of points representing known taxa by adding additional points corresponding to hypothetical ancestral taxa) for n taxa. The choice is made by ordering the possible solutions by length, where length is defined as the sum of the Manhattan distances between the nodes (points) in character space (Farris 1970). By analogy, parsimony in panbiogeography is a rule for choosing a subset of trees from the set of n possible undirected spanning trees for n points (Page 1990a).

To be informative, any statement of relationship between objects must be capable of prohibiting some relationships. If only one relationship is possible, then no information can contradict that relationship, and hence it is not testable. A spanning tree can be drawn between two points, but because the resulting tree is the only one possible, it is uninformative about biogeographic relationships. The addition of a third point allows for three possible spanning trees to be drawn (fig. 5-8). Choosing any one excludes the other two and, in a scientific sense (Popper 1972), that tree has empirical content. Similarly, in cladistics the simplest possible statement is a three-taxon Steiner tree, where one taxon is the root. Because there is only one possible Steiner tree for three points, the statement of relationship is uninformative. Adding a fourth taxon increases the possible Steiner trees to three, and an informative statement is now possible (fig. 5-8) (note that the "three-taxon" statement of cladistics is an implicit four-taxon statement; Page 1988).

For n localities or areas, there are n^{n-2} possible spanning trees. For two localities or areas there is a single possible spanning tree, for three localities or areas there are three, and for four localities or areas there are sixteen (fig. 5-9). Drawing any one track for three localities or areas excludes two other possibilities, so three localities or areas is the minimum number for which an informative track can be drawn using this method (Craw 1982). The presupposed superiority of cladistic methods (Platnick and Nelson 1984, Humphries 1985, 1992) over MST track methods in terms of the exclusion of possible relationships is consequently unsupported.

Nodal Analysis

A node is a geographic locality or distribution area where two or more tracks intersect or meet and corresponds to the graph theory concept of a node as a point of high connectivity (Page 1987, Henderson 1990). Nodes are of biogeographic interest because they represent localities or

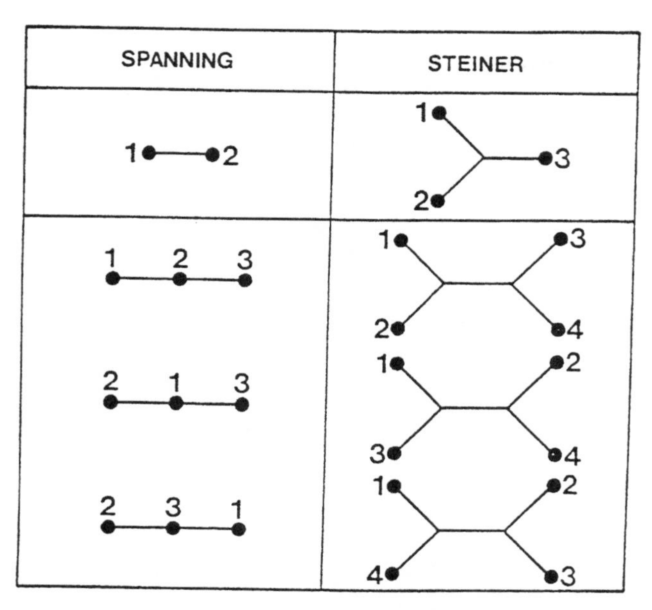

FIGURE 5-8. Informative statements in minimal spanning and Steiner trees. (Left) Addition of a third point allows for three possible spanning trees to be drawn, (right) addition of a fourth taxon allows three possible Steiner trees.

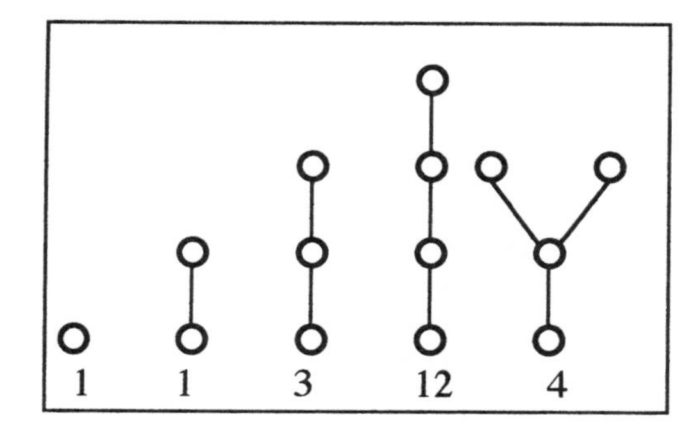

FIGURE 5-9. Number of possible minimal spanning trees from two localities or distribution areas (open circles) to four localities or distribution areas (from Page 1990: fig. 1; reprinted with permission of *Systematic Biology*).

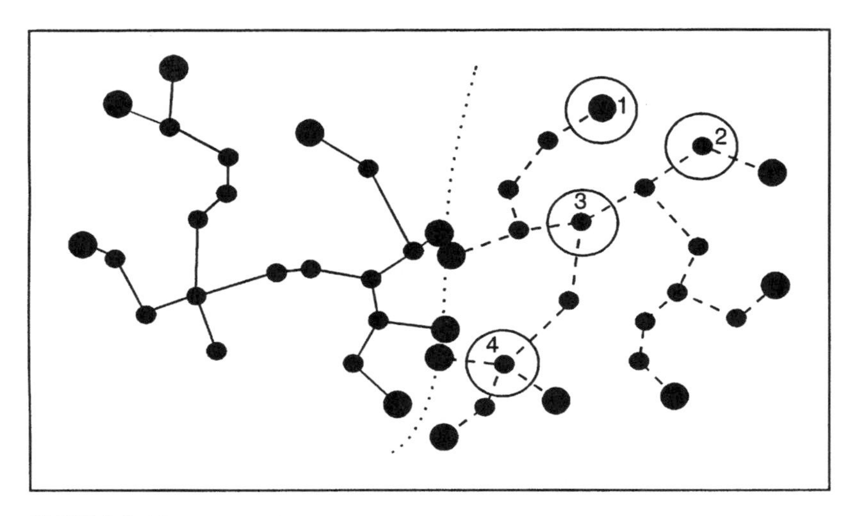

FIGURE 5-10. Minimal spanning trees and vertices of 1°, 2°, 3°, and 4° for two hypothetical related vicariant taxa (solid and dashed lines). A total of 16 1°, 18 2°, 8 3°, and 2 4° nodes are shared among the two taxa. An examples of each vertex degree is circled and labeled with the number of links, and 1° vertices are represented by larger symbols. Note that the highest density of 1° vertices is close to the boundary between the two taxa (dotted line) (redrawn from Henderson 1990: fig. 11c; reprinted with permission of SIR Publishing).

distribution areas with diverse biogeographic relationships in terms of geographic and phylogenetic affinities, because they correspond to the geographic and phylogenetic limits of taxa, and because they exhibit the presence of local endemics. Development of algorithms to draw tracks provides a foundation for quantitative analysis of nodes. This allows for the calculation of indices of nodal importance.

Localities connected by MSTs correspond to vertices in graph theory. Track vertices can be categorized by the number of links they share with other vertices. Terminal or endpoint vertices have only one connecting link with another vertex and are labeled 1° vertices. Vertices with two links with other vertices are 2° vertices, and these form the bulk of the internal part of any MST. Fewer vertices are 3° or more with links to other vertices (fig. 5-10).

These different levels of connection between vertices provide a method of nodal quantification according to the geographic limits of taxa. One-degree vertices tend to be distributed to the periphery of a distribution, and those localities where two or more taxa reach their distributional limits can be recognized by the presence of high densities of 1° vertices. For example, there are several vertices of 1° status around the peripheries of the two MSTs representing the hypothetical distribu-

FIGURE 5-11. Distribution of 1° vertices derived from vertex analysis of Trichoptera on the main islands of New Zealand. Size of shaded circles is proportional to the number of 1° vertices in each 10 x 10 km grid square (from Henderson 1990: fig. 13; reprinted with permission of SIR Publishing).

tions of two taxa in figure 5-10. Where these 1° vertices from both MSTs are in close proximity, their highest density will be close to the boundary between the two species (fig. 5-10).

This method allows nodes to be objectively defined for complex, overlapping distribution patterns. It has been applied to the New Zealand caddisfly (Trichoptera) fauna (Henderson 1990, 1991). Distribution of track vertices of 1° for 159 species was calculated (fig. 5-11). This distribution of vertices was tested against a null hypothesis of random distribution. Using the same location points, number of species, and

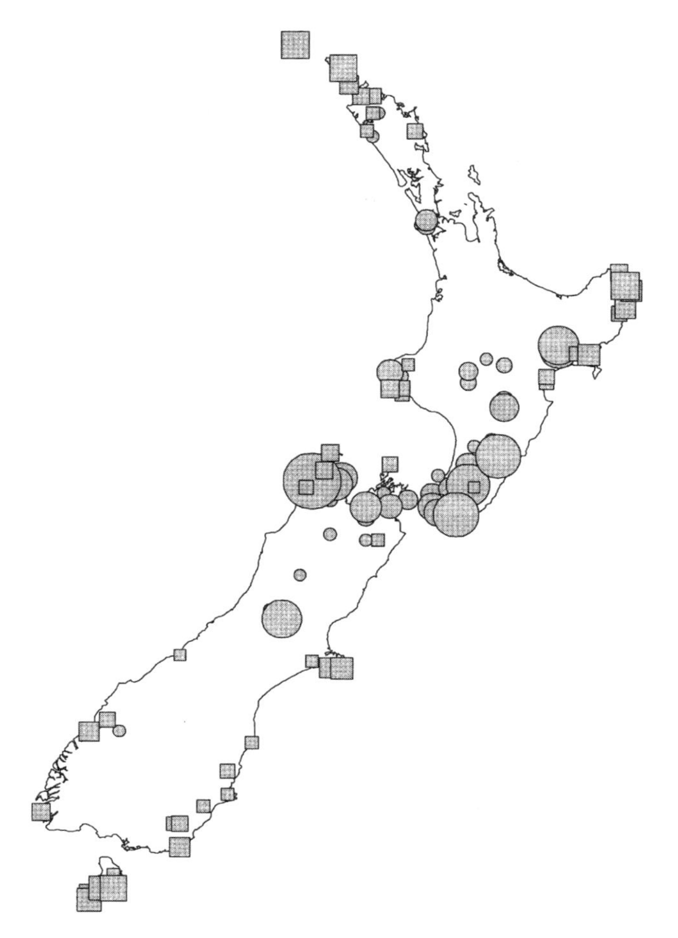

FIGURE 5-12. Nodes from figure 5-11 that were more intense than the 5% tail of the random distribution in a Monte Carlo simulation plotted as shaded circles, with diameter proportional to the excess of terminal vertices over the mean from the random model. Points where less than the expected number of terminal vertices are found are plotted as shaded squares. These correspond to antinodes (from Henderson 1991: fig. 11; reprinted with permission of *Australian Systematic Botany*).

records as in the initial study, a Monte Carlo approach was used to ascertain the statistical distribution of nodal intensity. This analysis identified many nodes more intense than the 5% tail of the random distribution. These were plotted on a map as circles with their diameters proportional to the excess of terminal vertices over the mean from the random model (fig. 5-12). Conversely, location points with fewer than the expected number of terminal vertices can be identified too. These

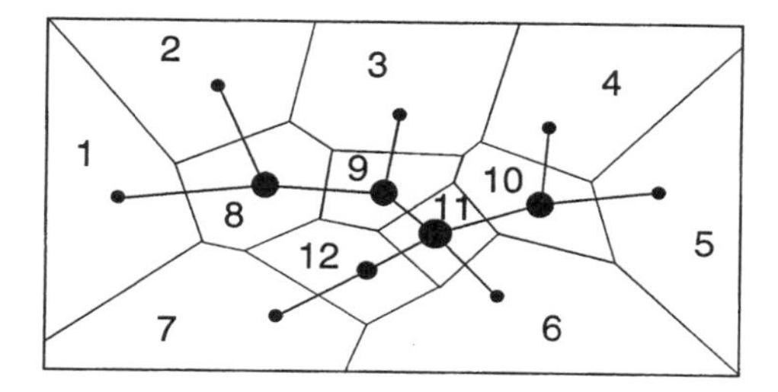

FIGURE 5-13. Recognition of vertices of high degree in a minimal spanning tree (MST) track for a hypothetical genus with 12 vicariant species. The vicariant species distributions are represented by the polygons numbered 1–12. Vertex degree in the MST linking the centers of each vicariant species range is emphasized by circle size (e.g., the smallest circles represent 1° vertices, the largest circle a 4° vertex) (redrawn from Henderson 1990: fig. 12).

antinodes represent significant levels of absent distributions, and these are also informative in biogeographic analysis (Heads 1990, Henderson 1991).

These methods can be applied to tracks of higher taxa (e.g., genera, families) and used to recognize nodes. Examination of 1° vertices of vicariant species tracks for species in a genus will tend to highlight the species boundaries, but not the overall nodal center for the group. However, an MST at the generic level, linking the centers of each vicariant species range, will include one or more vertices of higher degree (fig. 5-13). Vertices of high degree in tracks of higher taxa can also be recognized as nodes (Henderson 1990).

Compatibility Track Analysis

Compatibility track analysis combines the concept of distributional compatibility (Craw 1989, Connor 1988) with methods of generating random regular graphs (Wormwald 1984). In this method, individual tracks are treated as biogeographic hypotheses of locality or distribution area relationship. A track graph of a taxon's distribution (fig. 5-14) is treated as a vertex in a bipartite graph representing the relationship between localities or distribution areas which are or are not connected by that particular track (fig. 5-15). Individual tracks are regarded as being compatible only if they are the same in pairwise comparison or if one track is a subset of the other (i.e., tracks are either included within or replicated by one another). In the example given, tracks 1, 2, and 3 are all compatible with each other, whereas track 4 is incompatible with all three tracks.

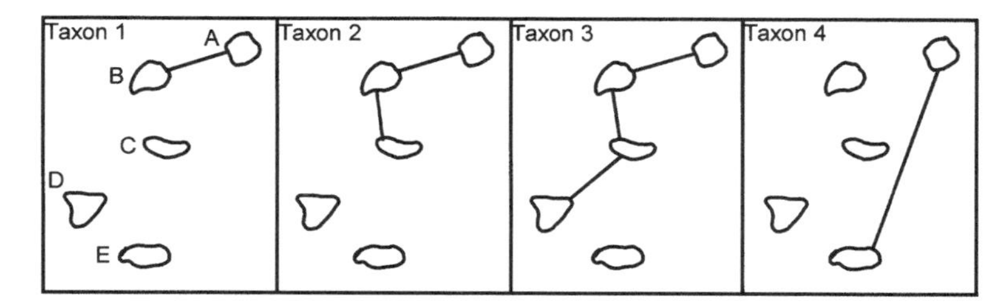

FIGURE 5-14. Individual tracks for hypothetical taxa 1–4 variously distributed in hypothetical distribution areas A–E.

By treating individual tracks in this fashion, it is possible to construct a locality/distribution x track matrix that can be analyzed for track compatibility by exploiting the analogy with the character compatibility approach to phylogenetic systematics (Meacham 1984). This matrix is an $r \times c$ matrix, where r, the rows, represent localities or distribution areas, and c, the columns, represent tracks. Each matrix entry, m_{ij} is 1 or 0 depending on whether track i is present or absent in location j. A compatibility analysis program (e.g., CLIQUE program in the PHYLIP package of Felsenstein 1993) is used to find the largest clique of compatible tracks.

The method involves finding a simple form of spanning tree linking localities or distribution areas. This tree is constructed from the largest clique of compatible distributions in a distributional compatibility matrix and is identified as a standard or generalized track. If more than one largest clique or several cliques of considerable size are found and these cliques are outside the constraints of what might be expected randomly, then a hypothesis of the existence of several standard tracks linking the localities or distribution areas in more than one way can be considered. Alternatively, the intersection (i.e., those tracks common to

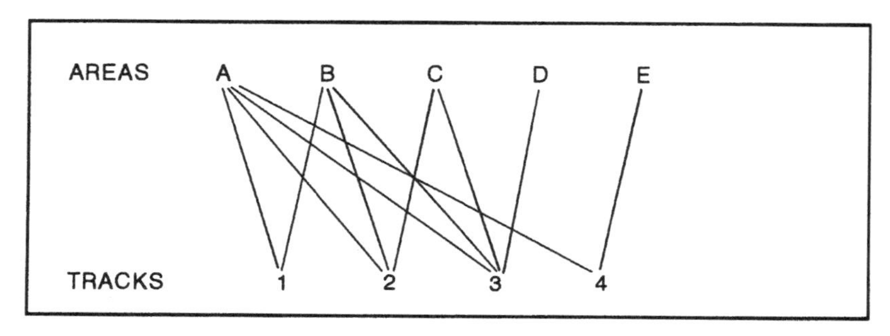

FIGURE 5-15. Tracks for taxa 1–4 variously connecting areas A–E in fig. 14 represented as a bipartite graph.

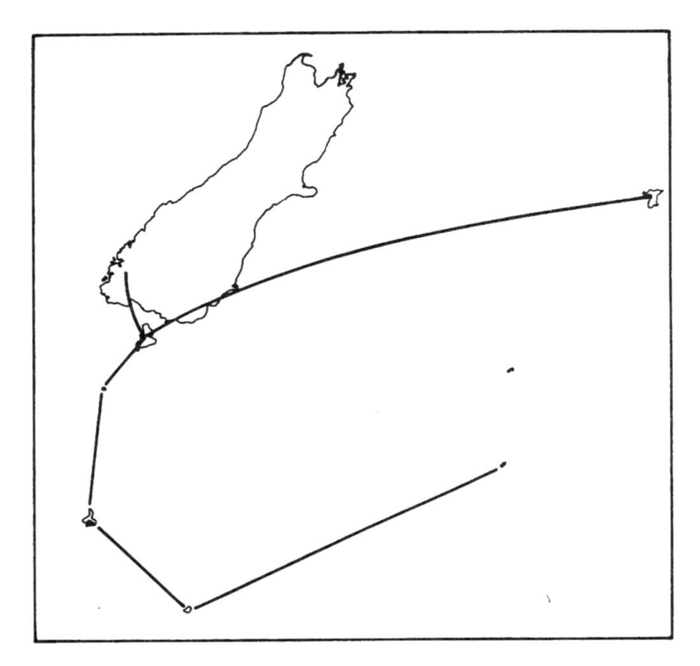

FIGURE 5-16. Chatham Islands–Southern New Zealand–subantarctic islands generalized or standard track based on the largest clique of 12 compatible tracks from an original sample of 18 tracks (from Craw 1989: fig. 15; reprinted with permission of *Systematic Biology*).

all the largest cliques) of the largest cliques can be identified as a standard track (Craw 1990).

This method was first applied in an analysis of the biogeography of insect and plant taxa disjunct between the southern New Zealand mainland and subantarctic islands, and the Chatham Islands (Craw 1989). Tracks for taxa with diverse means of dispersal ranging from wind- and bird-dispersed plants to flightless weevils belonged to the largest clique (12 out of 18 tracks) which was mapped as a standard track (fig. 5-16). Statistical hypothesis testing is possible with this method (Craw 1989, 1990) by using the BIPART algorithm (Wormwald 1984) to generate 1,000 or more equiprobable random matrices with given row and column sums. This algorithm allows random (null) incidence matrices to be generated within the constraints of given row and column sums. The biotic richness of localities or distribution areas and the frequencies of different taxa in the original data set are retained by these constraints. The percentage of randomly generated matrices, in which the largest clique size is as large or larger than the largest clique in the original data matrix, provides a statistical test of the level at which the largest clique attains significance.

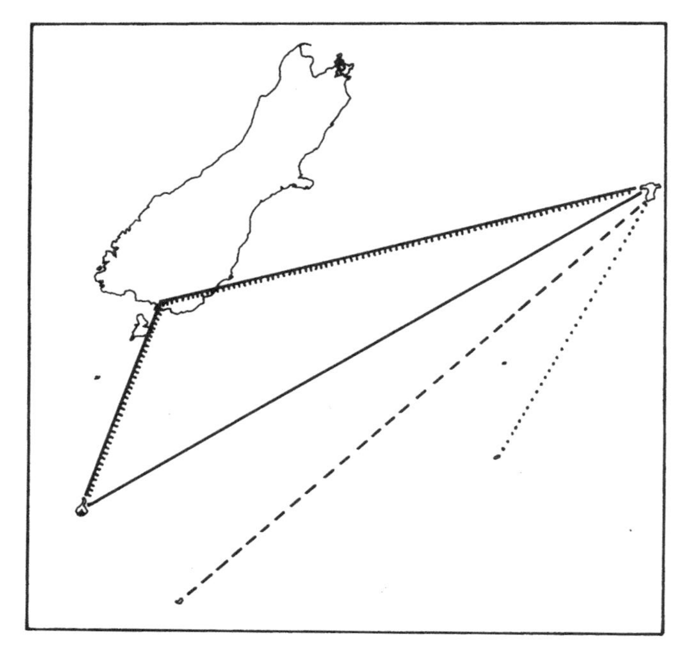

FIGURE 5-17. Examples of individual tracks for taxa variously distributed in the Chatham Islands and /or subantarctic islands and Southern New Zealand incompatible with the largest clique of compatible tracks mapped as a generalized track in figure 5-16 (from Craw 1989: fig. 14; reprinted with permission of *Systematic Biology*).

The null hypothesis was taken to be the probability of a random occurrence of a clique of size 12 or larger in a matrix of distribution areas x tracks of the same dimensions as the original data matrix. Four thousand random matrices were generated, and of these, 145 (3.6%) contained cliques of size 12 or larger, with $p<.04\%$, allowing for rejection of the null hypothesis. The standard track represented by the largest clique was interpreted as the trace of an ancestral biota that had been fragmented by vicariant events, although it was acknowledged that this track could reflect other factors such as a concordant dispersal pathway. Tracks incompatible with the standard track were also mapped (fig. 5-17), and the reasons for their incompatibility (e.g. chance dispersal, extinction) discussed on a case-by-case basis (Craw 1989, Lovis 1990).

Compatibility track analysis has been applied to 50 taxa (including ferns, angiosperms, crustaceans, and earthworms) distributed in the Falkland Islands, Tierra del Fuego, southern South America, and various subantarctic islands. A number of largest cliques were found, allowing for the recognition of five generalized tracks that intersected in the Falkland Islands and Tierra del Fuego. These two areas were identified as nodes on this basis (Morrone 1992). Distributional patterns of freshwater Decapoda (Crustacea: Malacostraca) in southern South America (Morrone and Lopretto 1994) and rhytirrhinine weevils (Coleoptera: Curculionidae) of the Andean provinces of South America (Morrone 1994) have also been analyzed with this method.

Grid Analysis and Main Massings

A main massing is a geographic localization of diversity for a taxon, whether taxonomic, ecological, genetic, morphological, or biochemical (Craw 1985). Grid analysis is a quantitative method for locating main massings (i.e., centers of taxic diversity and localized endemism of taxa). A grid appropriate to the geographical scale of the available distributional data is constructed. These distributional data are recorded in the geographic grids for each taxon and then superimposed to calculate total taxic density for each grid subdivision (fig. 5-18a). This method has been used to locate main massings for a variety of animal and plant taxa in New Zealand (fig. 5-18b) (Heads 1993, 1994a,c, 1997).

Grid analysis has been discussed in detail elsewhere (McAllister et al. 1986). McAllister et al. suggest that this method could be used to statistically test for correlations between distribution patterns and geological, geographic, and climatic features. This method could also be used as an alternative means of quantitative track analysis as proposed by McAllister et al. (1986:49):

> Tracing a straight line across the least gap between two related allopatric taxa (or their centroids) would permit quantification of Croizat's track analysis. . . . If the quadrats on track paths were scored with ones and these track path values were accumulated for each pair of taxa, then track paths could be numerically instead of subjectively identified and vicariance hypotheses about the tracks could be statistically tested.

Parsimony Analysis of Endemicity

Rosen (1984) first suggested parsimony analysis of endemicity (PAE) as an explicit means of discovering and analyzing endemic patterns of reef coral distribution and later developed this method further (Rosen 1988, Rosen and Smith 1988). Parsimony analysis of endemicity classifies localities or distribution areas by their shared taxa analyzed according to the most parsimonious solution. As Smith (1992:265) notes, "In effect it is a method of nesting panbiogeographical tracks into a hierarchical scheme."

Analytically PAE finds its rationale in the assumption that the occurrence of a particular taxon in a given locality or area can be interpreted

a

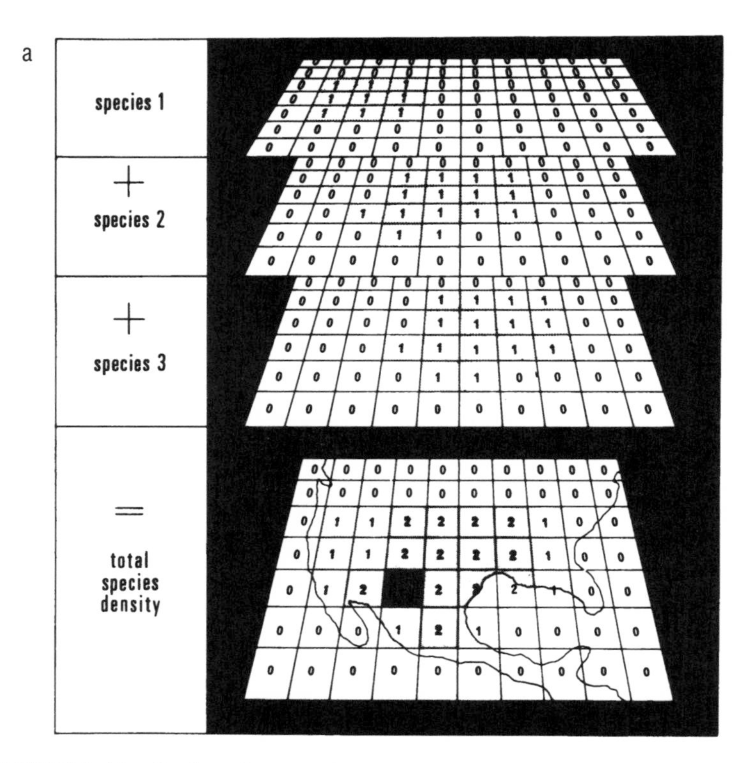

FIGURE 5-18. Grid analysis to determine geographical location of main massings. (a) Method of calculating total species (or other taxon) density or main massing. For each species (or other taxon), 1 represents presence and 0 represents the absence of the species (or other taxon) in a cell (in this case a degree square). The degree squares are summed down onto the total species (or other taxon) density map (from McAllister et al. 1986: fig. 2.1a; reprinted with permission of John Wiley & Sons, Inc.). (b) Application of this method to number of species and subspecific taxa of New Zealand *Parahebe* in each degree square (from Heads 1994a: fig. 4; reprinted with permission of *Botanical Journal of the Linnean Society*).

as a character, while the two character states are presence and absence (Crovello 1981). In PAE shared presences of taxa are treated like synapomorphies in cladistic systematics. From presence/absence matrices of distributional data for taxa from given localities or areas, locality or area cladograms can be generated. Cladograms of localities or areas are interpreted as expressing relationships between biotas, as represented at these localities or areas, with terminal dichotomies representing localities or areas most recently connected in space and time. Reversal and parallelism/convergence of character states are interpreted biogeographically as representing extinction and dispersal, respectively. Craw (1989)

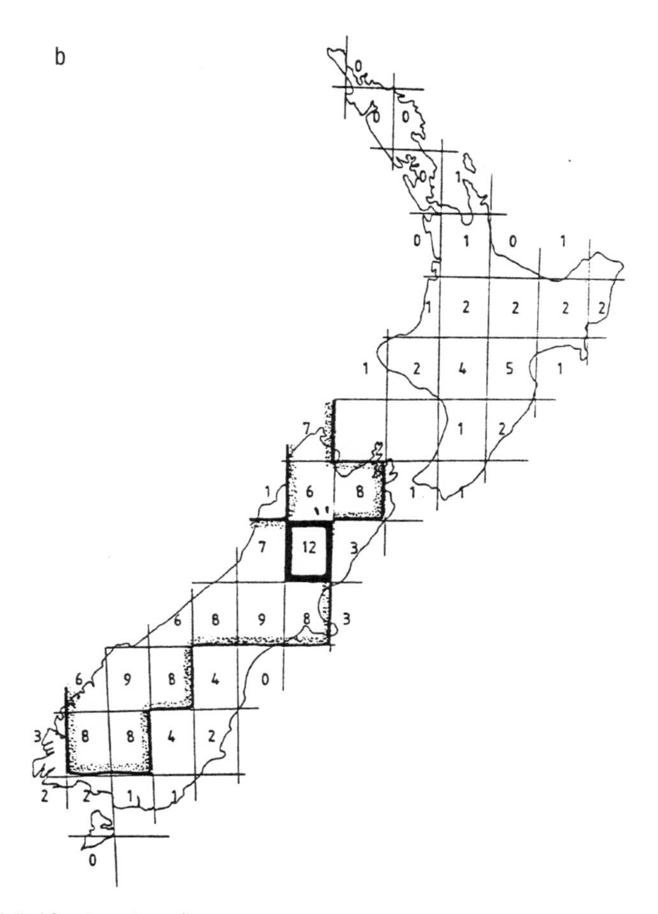

FIGURE 5-18. *(continued)*

has suggested modifying the approach when appropriate by treating putatively monophyletic clades (e.g., species groups, genera, families) as characters, and the various included taxa or combinations of taxa as alternate multistate characters of the localities or areas in which they occur (see also Cracraft 1991, Maldonado and Uriz 1995).

The most comprehensive application of this method is the study of Rosen and Smith (1988) on recent and lower Miocene corals and sea urchins and Eocene and late Cretaceous sea urchins. More recently, Smith (1992) used PAE in a study of sea urchin distribution in the Cenomian. Other PAE studies include Craw's (1989) on broad areas of endemism in New Zealand, Craw's (1990) on Southern Hemisphere disjunct families and genera of plants and animals, Myers's (1991) on Indo-Pacific and Caribbean gammaridean Amphipoda, Cracraft's (1991) on the Australian avifauna, and Maldonado and Uriz's (1995) on Mediterranean and Atlantic sponges.

5.4 Conclusions

This chapter has demonstrated that biogeographic line graphs (i.e., tracks) are a valid form of location-based representation in biogeography. Tracks, nodes, and main massings can be described effectively and defined empirically through application of a variety of quantitative techniques. Tracks, nodes, and main massings constitute empirical findings and are one basis for biogeographic analysis. Biogeographic hypotheses based on tracks can be tested and they are open to refutation. The track approach is a constructive method for investigation of the locality records of the geographic distribution of organisms, living and fossil, on a strictly comparative, factual, and statistical basis. Recently, Rainboth (1996:19) has noted that the "panbiogeographic method offers the best means for resolving certain reservations with current [biogeographic] methods." Similarly, Pohelmus (1996:64) has commented that "track analysis . . . would seem to be a reasonable tool [to use] in attempting to identify [island] arc-mediated distribution patterns." There is thus a growing recognition that panbiogeography can contribute to biogeographic analysis. Many more studies using the quantitative methods we have outlined above are now needed.

6

Toward a New Regional Biogeography

The Revival of Biogeographical Classification

From the first half of the nineteenth century to the present, the study of regions has largely defined the boundary of biogeography. The discipline occupied a comfortable niche in biology that assigned to it the integrative study of geographical distribution. Fulfilling this role, the principal task of biogeography (as phytogeography and zoogeography) was discovering and assembling facts about the geographical distribution of life. Whether such facts were geographic, palaeontologic, or systematic, the ultimate goal was a regional chorographic synthesis.

Traditionally, these syntheses proceeded by the time-honored practice of describing and listing floral and faunal elements and in the recognition and classification of floristic and faunal areas, provinces, and regions. Even before Darwin's (1859) *Origin of Species*, naturalists were intensely aware of regionalization in biotic distribution. Almost from the beginning of natural history, determined efforts were made to construct biogeographic classifications. A variety of independent classifications were developed during the nineteenth century, but the scientific and systematic study of different species in different areas goes back to Buffon's work in the eighteenth century on the differing mammals of the Old World and New World tropics (Nelson 1978, 1983). In the 1820s this observation was generalized by A. P. de Candolle to include a number of regions separated by barriers. De Candolle at first determined 20 botanical regions separated by barriers sufficient to impart isolation, but this was eventually increased to 38. This restatement of "Buffon's law" to include the existence of distinct regions now called "areas of endemism" was identified by Nelson (1978) as the birth of biogeography. Nelson (1978) suggests that the history of biogeography is the history of Buffon's law because biogeography is concerned with the nature and cause of biogeographic regions or areas of endemism.

Probably the most generally known biogeographic classification of areas is that of Wallace (1876) (fig. 6-1a) comprising only six major regions instead of the 20–38 proposed by de Candolle. Wallace believed that a similar classification by Sclater (1858) for birds (fig. 6-1b) was applicable to organisms in general (Nelson 1978). According to Nelson (1978), regional biogeographers were largely concerned with "Sclater's problem": how to determine the primary divisions of the earth's surface, with most approaches maintaining the six primary regions of Sclater and Wallace. The zoologist Huxley (1868), for example, emphasized latitudinal relationships and circumpolar unity (fig. 6-1c), whereas in botany, Engler (1882) proposed only four major areas (fig. 6-1d) later modified by Good (1964), especially for the Southern Hemisphere (fig. 6-1e). Wallace (1876) also provided subdivisions of the major biogeographic areas, and this tradition has continued to the present day, largely for terrestrial habitats, although recent efforts have also targeted marine ecosystems (Backus 1985, Stoddart 1992).

Alongside the development of historical biogeographic classifications were efforts to construct ecogeographic systems. This approach was advocated by Allen (1871) in recognition that adaptations to natural surroundings generally confine species within definite areas. These adaptive constraints could be mapped latitudinally in relation to circumpolar climatic zones, whereas longitudinal differences within a zone were affected by the presence of ocean barriers (Udvardy 1969). Subsequently, ecological areas or biomes were delineated by the correlation of plant or animal distribution with climatic, edaphic, and other environmental factors (Udvardy 1969, Cox and Moore 1993). Historical and ecological classifications are sometimes combined, with historical classification applied to continental or regional units, and ecological biomes as local subdivisions within each unit.

Udvardy (1969) concluded that there seemed to be about as much disagreement over regional boundaries among botanists as there was among zoogeographers. The general principles, concepts, and constructs of the Sclater-Wallace system, however, were widely accepted in Darwinian evolutionary biology, and they are still accepted today. A unifying theme was the existence of discrete areas of endemism that could be given definitive geographic boundaries—whether historical or ecological. Because these boundaries correspond with continental outlines, the introduction of plate tectonics did not threaten their integrity (Craw 1982) (fig. 6-1f). Nelson (1983) characterized Wallace's classification as

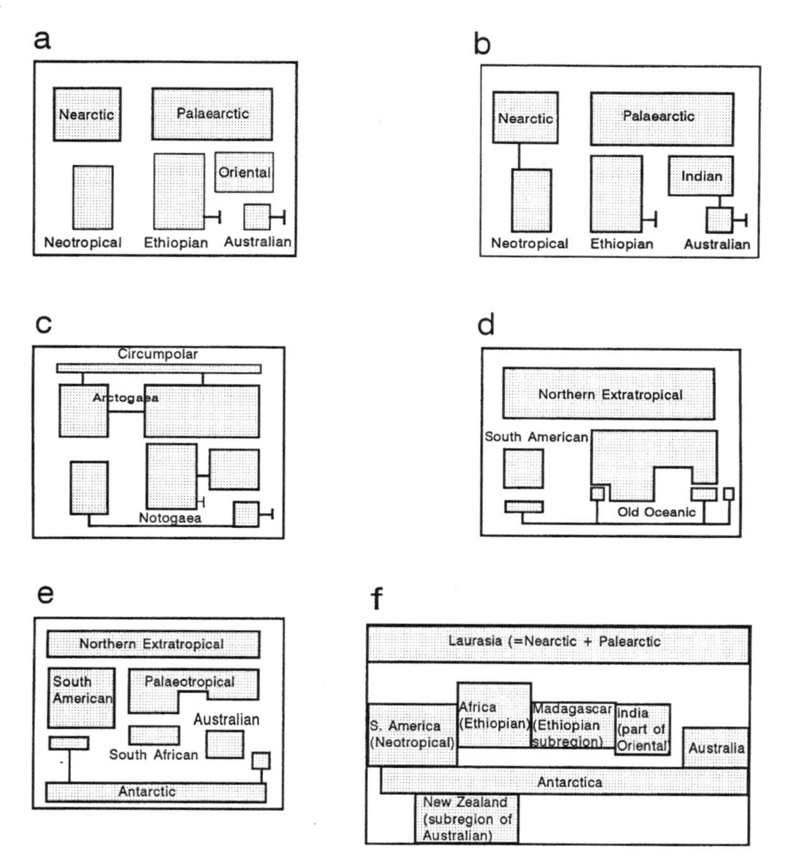

FIGURE 6-1. Schematic models of biogeographic classification systems: (a) Wallace (1876), (b) Sclater (1858), (c) Huxley (1868), (d) Engler (1882), and (e) Good (1964) (after Udvardy 1969: figs. 5-9). In each of these models the present major landmass areas constitute the focus for delineation of biogeographic boundaries. (f) Congruence of Wallacean classification and continental drift (after Craw 1982: fig. 3; reprinted with permission of *Systematic Biology*). Despite the mobilism injected by continental drift, the overall integrity of the Wallacean areas is maintained.

an artificial construct explicitly based on criteria such as convenience and utility. The Sclater-Wallace system (including biome–ecological methods) also fails to meet the fundamental requirement of natural biological classification—the designation of explicit, empirical statements of relationship that allow for falsification. To address this issue, a concept of geographic congruence homologous with that of phylogenetic congruence in organisms is needed (cf. Nelson 1983).

6.1 Spatial Logic and Homology in Biogeography and the Earth Sciences

Spatial Logic

Alfred Wegener proposed his theory of continental drift in the early decades of this century on the basis of biological and geological similarities between the disjunct landmasses, particularly those of the Southern Hemisphere. These "homologies" included fossil resemblances, stratigraphic similarities, continental fits, and paleoclimatic indicators (Oreskes 1988). Later authors used line graphs, equivalent to biogeographic tracks, to trace such similarities in geological features across ocean basins (e.g., Haddock 1938, Holmes 1944). They argued that these stratigraphic and tectonic correspondences could not be merely fortuitous and that they were best explained by former contiguity between the continents before they drifted apart. Recently, Storey (1995) has used what equates to spanning tree line graphs (and in some cases minimal spanning trees) to illustrate the most likely connections between flood basalt provinces in the former Gondwana continental fragments (fig. 6-2).

These usages of geological track graphs are examples of what has been termed "spatial logic." Spatial logic is defined as the acceptance of "morphology, spatial distribution, and spatial association as primary evidence of earth processes that must be tested through process-oriented research" (Dobson 1992:187). Spatial patterns and associations are clearly an important part of both biogeographical and geological research. An integration of biogeography and geology may be possible around a combination of the comparative concept of homology and the method of spatial logic, and this may be significant for the field of biogeographical classification.

Track Homology

To establish hypotheses of congruent or incongruent distributions, geographic comparison of tracks requires a concept of spatial homology to provide a diagnostic character by which individual tracks can be compared and contrasted. This approach is different from simply comparing distributions for patterns of geographic overlap, congruence, or coincidence. Spatial homology is designated by a diagnostic character termed the "baseline."

For example, consider a related group of four vicariant genera (*a*, *b*, *c*, *d*) belonging to family *X*, distributed in northeastern South America (*a*), West (*b*), Central (*c*) and East Africa (*d*), and therefore disjunct across the Atlantic Ocean (fig. 6-3a). The track graph for this group crosses the Atlantic Ocean, and they are described as a provisional conjecture as having an Atlantic Ocean baseline. Another group of vicariant genera (*e*, *f*,

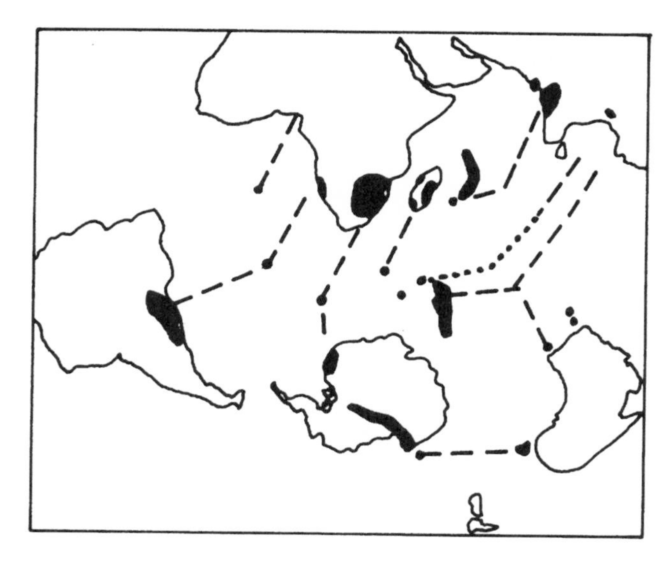

FIGURE 6-2. Map of Southern Hemisphere landmasses and India illustrating the most likely connections between flood basalts (after Storey 1995: fig. 4; reprinted with permission of *Nature*). The representation of these geological connections equate to spanning tree graphs, and inferences are comparable to that of biogeographical tracks.

g, *h*) belonging to family *Y*, and distributed in West (*e*), Central (*f*) and East (*g*) Africa, and Southeast Asia (*h*) are disjunct across the Indian Ocean (fig. 6-3b). They are assigned an Indian Ocean baseline. A third group of three vicariant genera (*i*, *j*, *k*) belonging to family *Z* are distributed in central South America (*i*), southeastern South America (*j*), and southwest Africa (*k*), and share an Atlantic Ocean baseline with the genera belonging to family *X* (fig. 6-3c). These two groups (genera in families *X* and *Z*) are regarded as biogeographic homologues, although they display no geographic overlap or congruence, because they share the same baseline.

Origin and Implications of the Baseline Concept

The baseline concept was introduced by Croizat (1952), who used it as a working hypothesis in track analysis to designate a means by which any transoceanic disjunctions in a taxon's distribution could be linked in a way consistent with the whole distribution according to the principle of parsimony (Croizat 1968a, 1971). A baseline represented a "center of origin" for the ancestors with subsequent migrations of a particular group extending the range to its current limits over past geographic

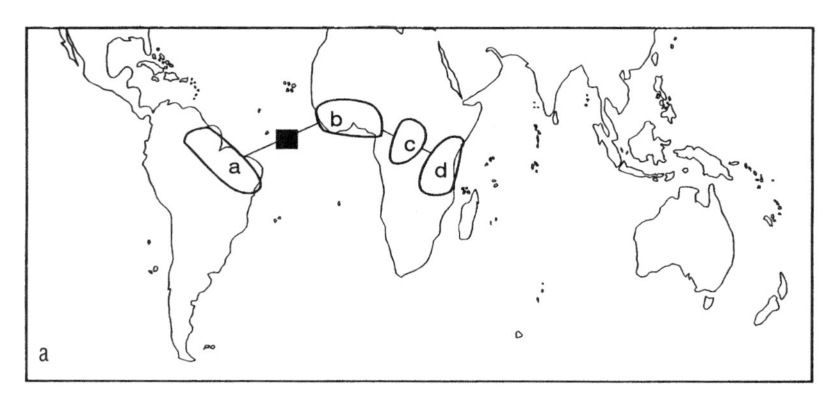

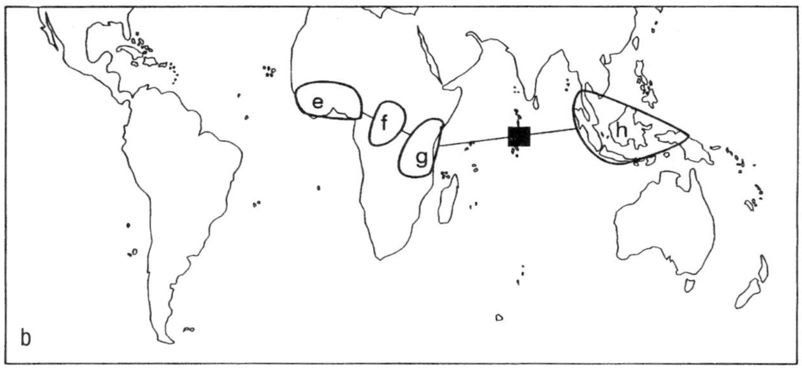

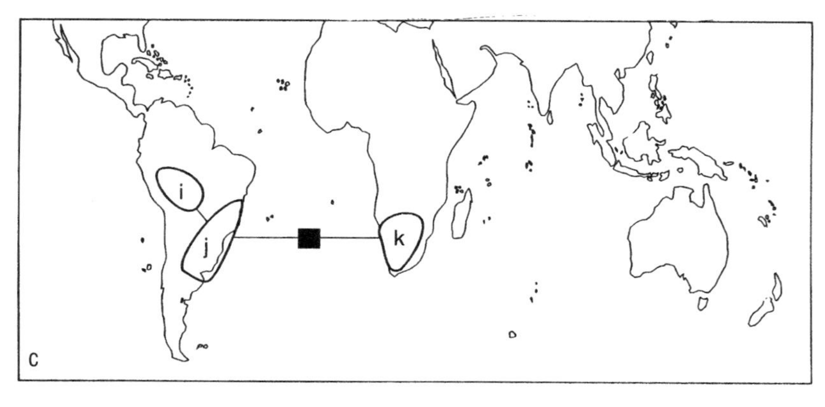

FIGURE 6-3. Concept of the baseline. (a) Hypothetical example of four related vicariant genera (*a, b, c, d*) belonging to family *X* with an Atlantic Ocean baseline; (b) hypothetical example of four related vicariant genera (*e, f, g, h*) belonging to family *Y* with an Indian Ocean baseline; (c) hypothetical example of three related vicariant genera (*i, j, k*) belonging to family *Z* with an Atlantic Ocean baseline.

landscapes (e.g., drifting and rifting continents, land bridges, or island chains) that originally connected the present disjunction. This baseline identification depended on a criterion of correspondence between the main massings in relation to the disjunct distributions within the range of the taxon. These centers of origin/baselines were recognized as ancestral areas located on a historical landscape that no longer existed. For example, the center of dispersal and "probable place of origin" of eyebrights (*Euphrasia*: Scrophulariaceae) was considered to lie somewhere in what is now the South Pacific. Tracks were thus oriented from this baseline, indicating a direction for migration out of the ancestral center not as a casual process, but as one mediated by past geographic, geological, and climatic conditions (fig. 6-4a).

This baseline concept represented a provisional conjecture on the spatial proximity of taxa with respect to a former ancestral range and acted as a reference point for interpreting the phylogenetic representation of characters within and between taxa. Thus, the presence of similar patterns in the variation of individual characters in disjunct taxa could be recognized as being derived from a common ancestor in the same way that individual species in a lineage share similarity through descent. Baselines provided an analytical concept for explaining the phylogenetic resemblance of taxa such as the same subspecies of *Dicrurus paradiseus* on both sides of the Bay of Bengal, or parallel north/south clines in spore development of ferns (*Schizaea* spp.: Schizaeaceae) on either side of the Pacific (fig. 6-4b).

A baseline is not a means to fix a precise center of origin, but a locus around which to rationalize explanations of geographic distribution. Once the baseline for a taxon is assigned to a particular geographic sector of the earth, it constrains the possibilities that may be considered as to how and why taxa or characters are distributed. If the Pacific is the main axis of distribution for a taxon, for example, the *Schizaea* species cited above, biogeographic discussion need not speculate on whether these ferns are "American" elements in Melanesia or "Melanesian" elements in America, or whether this fern may or may not have migrated from some other sector of the globe such as the Indian or Atlantic Ocean basins (Croizat 1968a).

Ocean and sea basins are the prime focus for major baselines because these basins constitute the geographic "anomaly" defining global biogeographic problems. An initial issue of interest for biogeographers is understanding why distributions are associated with particular ocean basins—why, for example, a distribution is associated with the Indian Ocean basin rather than with the Pacific Ocean or vice versa. While recognition of an Indian Ocean baseline for a particular group does not demonstrate an Indian Ocean origin for that group, it does illustrate that its distribution pattern is compatible with such an origin.

Lands and seas over and through which organisms may have moved

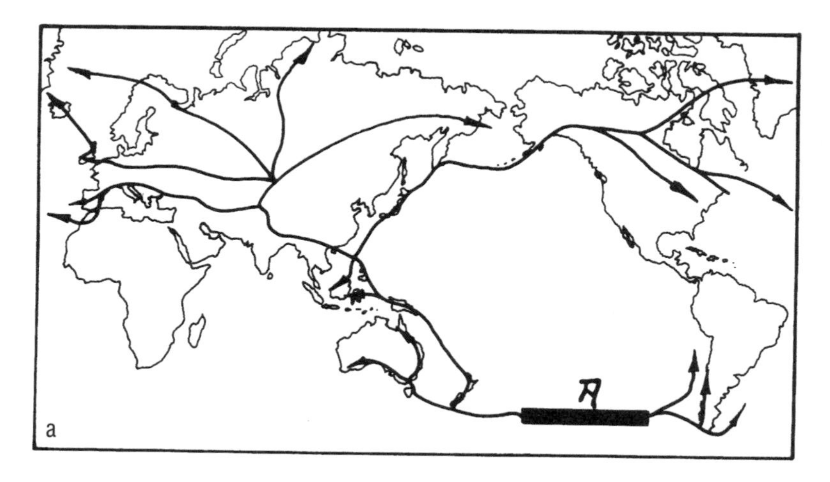

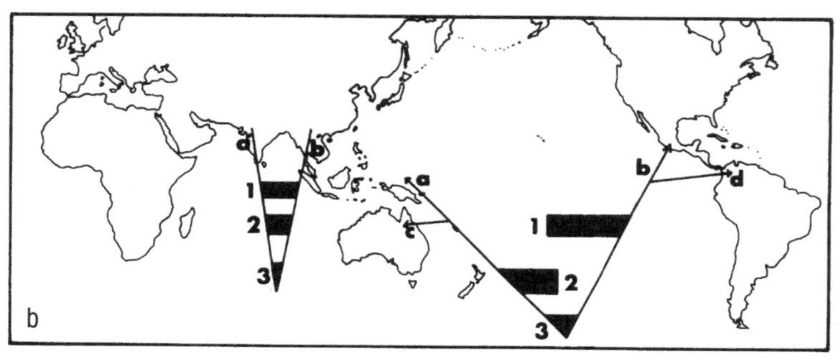

FIGURE 6-4. Origins of the baseline concept. (a) As a means by which transoceanic disjunctions in a taxon's distribution (in this case *Euphrasia*) can be bridged in a way consistent with the whole distribution according to the principle of parsimony (after Croizat 1952: fig. 11); (b) as an analytical concept for explaining the phylogenetic resemblance of taxa on either side of an ocean basin (from Croizat 1968a: fig. 41; reprinted with permission of Istituto Botanica, Pavia).

during periods of mobility do not necessarily correspond with the lands and seas where they now occur. The designation of an Indian and/or South Atlantic Ocean baseline correlates with popular plate tectonic models requiring a fragmentation and relocation of a former Gondwana supercontinent. These ocean baselines imply that the actual lands or seas over which organisms have spread include not only those where they now occur but also those that no longer exist due to geological change and upheaval. The present location of organisms is the result, not the origin, of their biogeographic history.

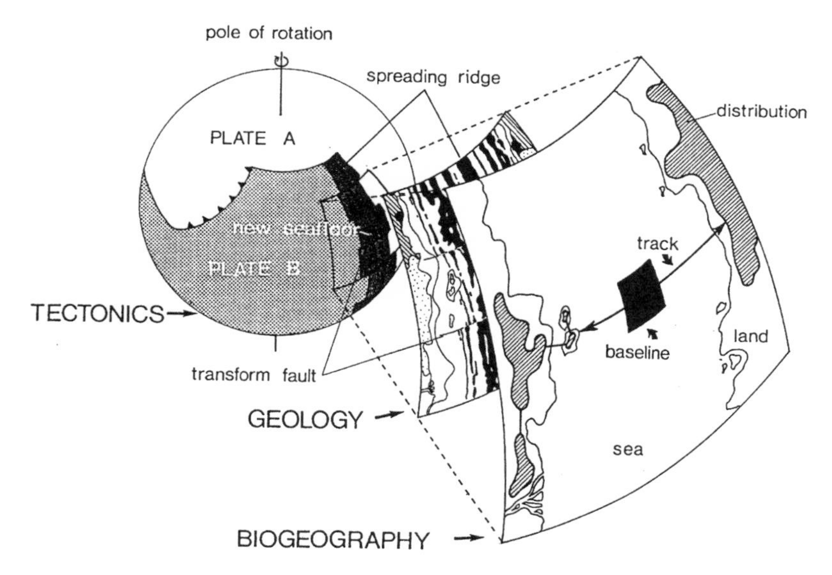

FIGURE 6-5. The baseline of a track reinterpreted in terms of the crossing of a major tectonic feature (from Craw and Page 1988: fig. 5; reprinted with permission of John Wiley & Sons).

Baselines, tracks, and tectonics

These considerations of the location of track baselines in ocean basins emphasized the incongruence between current geography and biological distribution and suggested the possibility of establishing a relationship between earth history and those distributions by matching or correlating biogeographic, geological, and paleogeographical patterns and processes. Distribution patterns and track geometries can be readily correlated with geological and geomorphological features and paleogeographical models (see chapter 2), but the concepts of track and baseline can be integrated as well with tectonic processes. This allows for the conversion of a track analysis into a dynamic explanation of geographic distribution in terms of biogeographical and geological history (Craw and Page 1988).

Midocean spreading ridges between tectonic plates can be visualized as resulting from the relative rotation between two plates. This relative motion is a rotation around a pole. By drawing tracks for taxa with transoceanic biogeographic relationships on maps reprojected using poles of rotation for the intervening plates, it can be shown that the biogeographic tracks are approximately parallel to one another and to the transform faults and orthogonal to the spreading ridges. The original idea of a track baseline as an ocean or sea basin is modified so that the baseline of a track is where that track crosses a major tectonic feature (fig. 6-5).

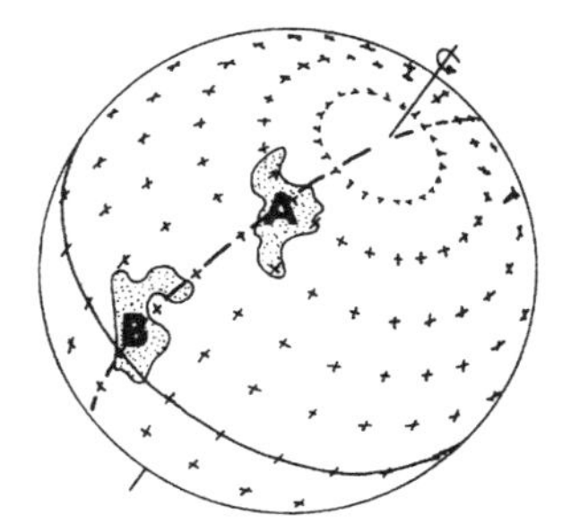

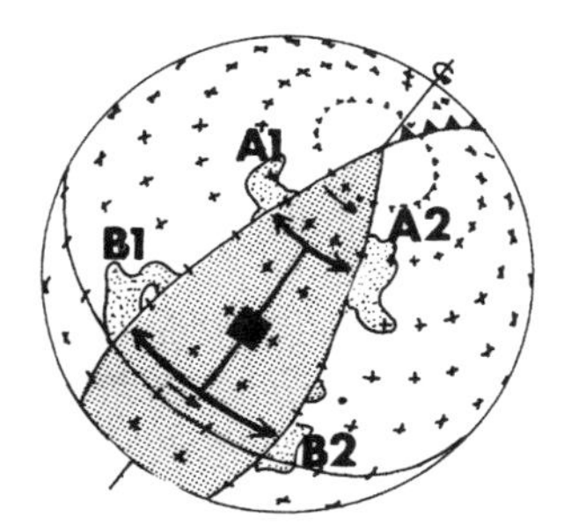

FIGURE 6-6. The relationship between track and plate tectonic geometry and baselines. (Left) Two groups of organisms (A and B) distributed across a future spreading ridge; (right) populations of the two groups of organisms (A1 and A2, B1 and B2) become disjunct as the spreading ridge develops. Both disjunctions can be related to the same spreading event so that the tracks linking the disjunct populations of each group of organisms can be said to share the same baseline (from Page 1990a: fig. 11; reprinted with permission of SIR Publishing).

This approach also explains why taxa with incongruent or noncoincident geographic distributions can be considered as biogeographically homologous. Consider two groups of organisms that are distributed across a future spreading ridge (fig. 6-6a). As the ridge begins to develop and sea-floor spreading occurs, the two plates on the globe rotate with respect to each other around the pole of plate rotation, and the two groups of organisms are now disjunct (fig. 6-6b). Both disjunctions can be related to the same vicariant, tectonic event, so the two tracks linking the different groups can be considered to share the same baseline. In addition, the two tracks are concentric on the pole of plate rotation, which is the same relationship shown by transform faults in the ocean and sea floors (Page 1990a).

6.2 Case Study: Transpacific Tracks and the Geological History of the Americas

The biogeography of the American continents is characterized by two major suites of standard track relationships: an eastern trending set with Atlantic Ocean baselines (i.e., a suite of transatlantic tracks for American taxa that are most closely related to taxa in Europe and Africa; see, for example, chapter 3), and a western trending set with Pacific Ocean baselines (i.e., a suite of transpacific tracks for American taxa with their nearest relatives in or across the Pacific, in eastern and southeastern Asia and Australasia). Taxa with distributional main massings in the west or the east of the American continents are components of transpacific or transatlantic tracks, respectively. Transatlantic standard tracks are well known, and along with trans-Indian Ocean and transantarctic tracks fit the conventional scheme of Pangean breakup. But this original bunching of all the continents around Africa, closing the Atlantic and Indian Oceans, greatly increases the width of the proto-Pacific Ocean (Panthalassa), and this is incompatible with transpacific biotic relationships (Wulff 1943, Skottsberg 1956, Croizat 1958, Nelson 1985, Craw and Page 1988, Sluys 1995).

Transpacific biogeographic relationships are sometimes explained as the result of differential extinction or as the consequence of migration around or across the Pacific basin (Eskov and Golovatch 1986, Cox 1990). But transpacific affinities and tracks are known from a wide variety of fossil and living organisms of freshwater, marine, and terrestrial habitats. They are as commonplace as the better known transatlantic and trans-Indian Ocean tracks, over which there is less controversy concerning their relationship to tectonic events in the respective ocean basins they cross. Twenty-four taxa of marine algae and fifteen terrestrial taxa comprise components of a transpacific Australasian–southern South America–western North America track (Chin et al. 1991). Five groups of fossil freshwater teleost fishes from the Eocene Green River formation of Wyoming, Colorado, and Utah in western North America feature transpacific relationships of an Australasian affinity (Grande 1985, 1989), and a close relationship between freshwater aspinine cyprinid fish taxa of western North America and East Asia (China and Japan) has been noted (Howes 1984). Ten fossil and seventeen living groups of western North American freshwater Mollusca exhibit transpacific affinities and tracks (Taylor 1988). Transpacific relationships have also been described for a diverse range of other taxa including butterflies and moths (Grehan 1991a), relict weevil taxa (Morrone 1996), earthworms (Sims 1980), keropelatid flies (Matile 1990), land planarians (Sluys 1995), land snails (Miller and Naranjo-Garcia 1991), the angiosperm plant family Portulacaceae (Carolin 1987), and aroid plants (Araceae) (Mayo 1993).

If these contrasting transatlantic and transpacific tracks are the result of the distribution of ancestral biota upon former landmasses in the Atlantic and Pacific, the American continent should have a composite geological origin rather than being solely the product of fragmentation after formation of the Atlantic as required in classical Pangaean theory (fig. 6-7). Formulated within a continental drift paradigm, this panbiogeographic model of the Americas represented a novel development when first proposed by Croizat (1958, 1961). In contrast with earlier geological or biogeographic proposals, Croizat suggested that geologically and biologically the continent is divided into eastern and western, rather than northern and southern, sectors (see Craw and Weston 1984, Craw and Page 1988).

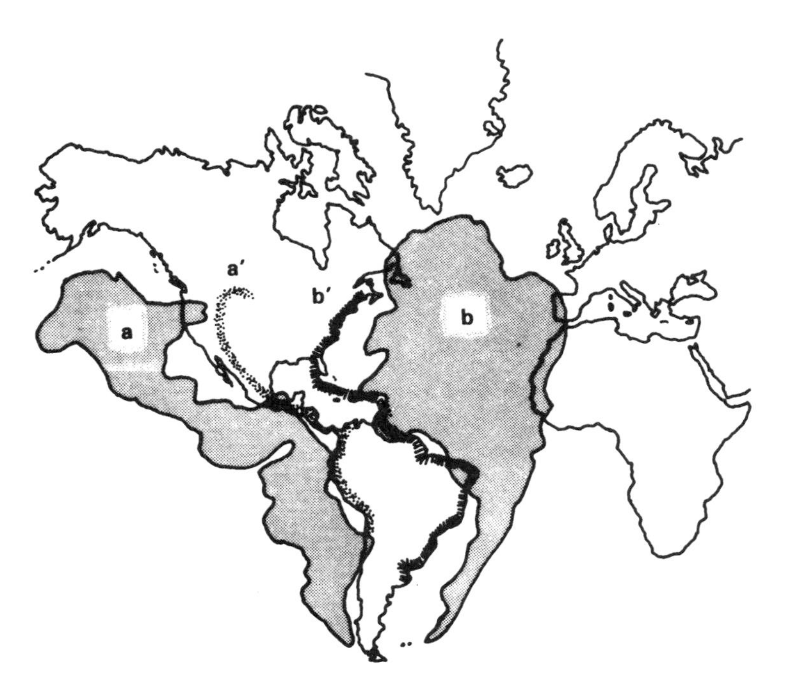

FIGURE 6-7. Model of a composite geological and biogeographical origin of the Americas with respect to two "halves" (a and b) that drifted together and accreted or collided to form the current geography. Organisms such as *Halenia* (Gentianaceae) were dispersed on the western half (a); others such as *Drosera* (Droseraceae) were dispersed to the east (b) and their respective modern ranges to the west (a′) and east (b′) continue to represent their former massings (after Croizat 1961: fig. 8).

The Pacific origin of western components of the Americas conflicts with conventional drift and plate tectonic theory (which allows only for an Atlantic derivation), but geologists now also recognize the allochthonous (exotic) origin of the numerous terranes found in the Cordilleran mountain systems of western North and South America (e.g., Feininger 1987, Jones 1990, Monger 1993, Mann 1995). On the other side of the Pacific basin microcontinents, island arcs and oceanic crust have been accreting along the Eurasian margins since early Paleozoic time (Maruyama et al. 1989). In particular, the space–time distribution of transpacific tracks points to a close association between the dispersal and accretion of island arcs, oceanic crust, and microcontinental terranes and the biogeography of lands bordering the Pacific Ocean and the former Tethys Sea. For instance, consider the transpacific tracks of a living angiosperm group *Coriaria* (Coriariaceae) along with that of a Triassic bivalve (*Monotis*). These two groups share a Pacific Ocean baseline, even

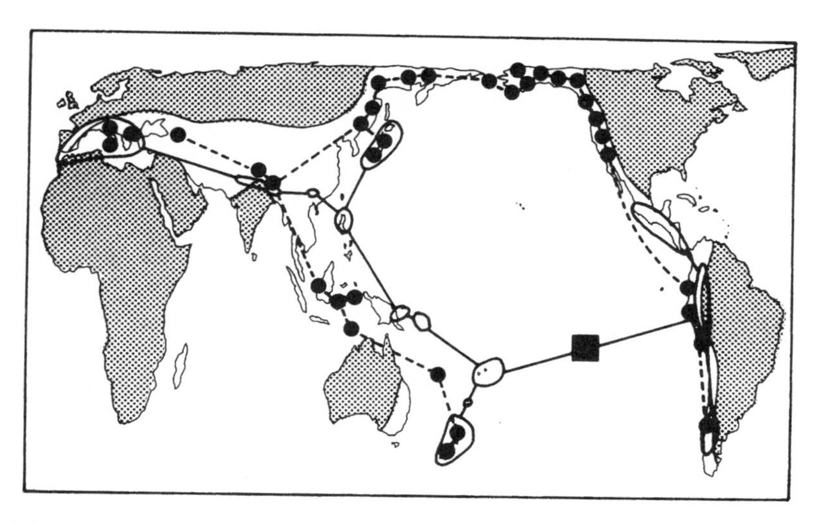

FIGURE 6-8. Distributions, tracks, and baselines (square) of the angiosperm *Coriaria* (distribution as solid lines, track as solid lines) and the Triassic bivalve *Monotis* (distribution as filled circles, track as dashed lines). Generalized extent of accreted terranes is shown as untextured surface. Note the geographic congruence between the fossil and living organisms and the geographic areas composed of accreted terranes (after Heads 1990: fig. 19, Grehan 1994: fig. 8).

though *Monotis* is represented in high northern latitudes, while *Coriaria* is not. Note also the spatial congruence of these two groups with the geography of accreted terranes (fig. 6-8). Other correlations of Pacific Rim distributions with allochthonous terrane boundaries include the frog genus *Ascaphus* in western North America (fig. 6-9a), Tertiary flora in Alaska (fig. 6-9b), Permian, Triassic, and Jurassic Tethyan fauna in North America (fig. 6-9c), bolitoglossine salamanders in northern South America (fig. 6-9d), and cyprinid fishes in East and Central Asia (Howes 1984).

The panbiogeographic model divides the Americas into two halves, whereas terrane models emphasize a number of "microcontinental" blocks and island arcs in Panthalassa (= proto-Pacific Ocean). These views are not unreconcilable because the panbiogeographic model was not concerned with specific geological detail, and some models (e.g., Roure and Sosson 1986, Lambert and Chamberlain 1988, Lapierre et al. 1990) envisage many of the suspect terranes amalgamated into microcontinental collages before colliding with the North American craton. In particular, the microcontinent of Cordilleria (Chamberlain and Lambert 1985, Lambert and Chamberlain 1988) bears a striking similarity to Croizat's (1958) northwestern North American region with predominantly transpacific biogeographic affinities (fig. 6-10). Some of these ac-

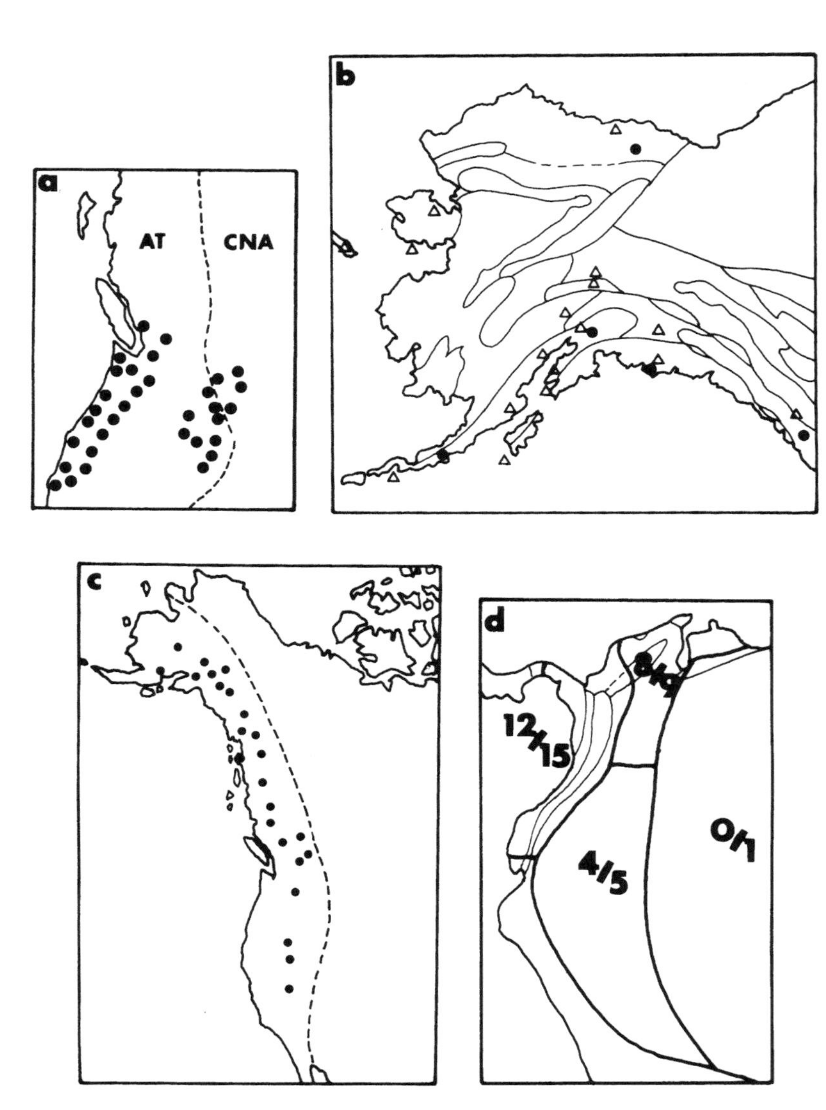

FIGURE 6-9. Tectonic correlations between terrane boundaries and continental distributions. (a) Distribution of the primitive frog genus *Ascaphus* and the boundary between the North American Craton and suspect Cordilleran terranes. (Dashed line) Boundary between allochthonous terranes (AT) and cratonic North America (CNA) (after Craw 1985: fig. 4). (b) Distribution of Tertiary floras in relation to the margins of identified allochthonous terranes in Alaska. (Light lines) Terrane margins; (filled circles) Paleocene and Eocene floras; (open triangles) Oligocene or later floras. All Paleogene floras except of the Sagavanirktok on the north slope lie directly on recently accreted terranes (after Tiffney 1985: map 1). (c) Permian, Triassic, and Jurassic Tethyan fauna (filled circles) in relation to the western limit of the North America craton (dashed line) (after Hallam 1986: fig. 2). (d) Bolitoglossine salamanders in northern South America. (Fine lines) Tectonostratigraphic terranes; (heavy lines) areas of endemism; ratios refer to number of endemic species/total species (after Hendrickson 1986: fig. 1).

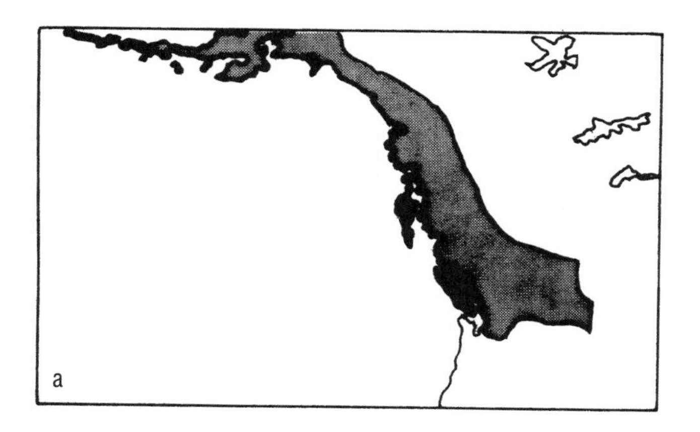

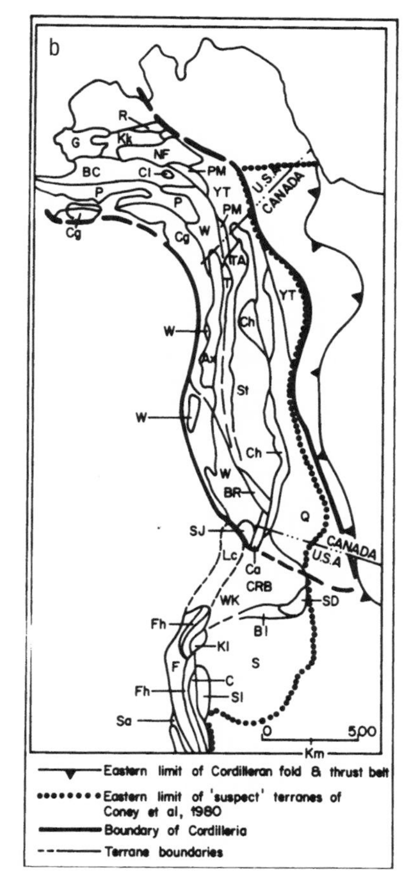

FIGURE 6-10. Similarity between (a) Croizat's (1958: fig. 56) definition of northwestern North America (stipple, upper left) involved in transpacific biogeographic relations and (b) Chamberlain and Lambert's (1985: fig. 4; reprinted with permission of *Nature*) microcontinent of Cordilleria.

FIGURE 6-11. Triassic latitudinal positions of North and South American suspect terranes proposed by Tozer (1982) (shaded) correlated with the Pacific "fragment" from Croizat's (1961) continental drift model (area enclosed by a solid line). This correspondence reflects a concordance between the biogeographic model and more recent proposals for Pacific origins of suspect terranes by tectonic theorists.

creted terranes are postulated to have traveled extensive distances before becoming part of North America (Ross and Ross 1985) and to have been originally positioned thousands of kilometers to the west of the North American craton (Belasky and Runnegar 1994). Triassic latitudinal positions of North and South American suspect terranes proposed by Tozer (1982) show a substantial correspondence with the Pacific fragment in the panbiogeographic model (fig. 6-11).

6.3 Conclusions

As shown in this and previous chapters, standard tracks, recognized through application of the panbiogeographic method, link areas now separated by vast expanses of ocean and sea. Portions of continents, large islands, and archipelagos are related to one another by tracks that

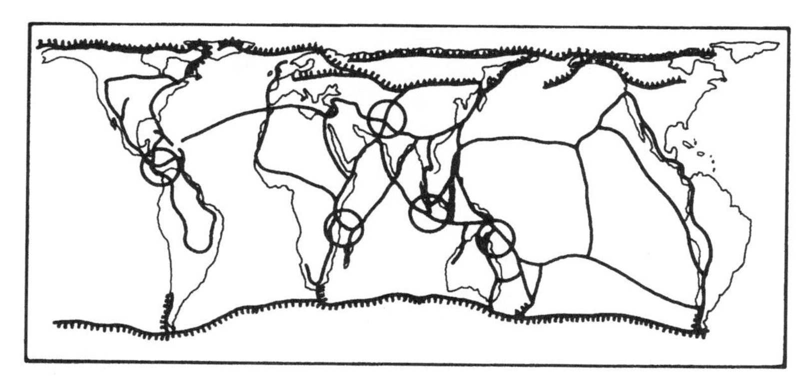

FIGURE 6-12. Global biogeographic track relationships. (Solid lines) Standard or generalized tracks (boreal and Austral tracks distinguished by hatching); (circles) five major biogeographic nodes (gates) at the intersection of several standard tracks. This panbiogeographic representation of life simultaneously allows more than one biogeographic relationship between any single de Candollean or Wallacean area (after Croizat 1958: fig. 259).

span ocean and sea basins. The panbiogeographic classification is based on this complex pattern of track relationships among taxa occupying different parts of the globe (fig. 6-12). This biological track "internet" simultaneously allows more than one biogeographic relationship between any single de Candollean or Wallacean area. The tracks do not intersect with any particular area as such, but with each other at nodes or gates. Here the biogeographic relationships are defined by different spatial homologies involving unique geographical and geological features of the earth (Craw 1983). This approach is significant for its lack of a "primary trunk and root" holding these features together as in earlier classifications (Nelson 1983).

The major biogeographical regions are not those of Sclater and Wallace, but regions that no one had thought of—the modern ocean basins (Croizat 1958, 1961, 1964; Nelson 1978, 1985; Craw 1982). The new classification system was formalized (Craw 1988, Craw and Page 1988, Parenti 1991) as a system of ocean basin relationships where the limits of geography are relative to the biogeographic pattern (fig. 6-13). In this panbiogeographic classification, there are five major regions, each represented by a major ocean basin designating a specific sector of the earth that identifies a particular spatiotemporal history for components of the standard tracks that cross that basin.

In this classification system, the Wallacean regions constitute artificial conglomerates of geology and biology. Each de Candollean or Wallacean region or area is a composite of contrasting biogeographic histories that amalgamated by the fusion of different biological and geological landscapes. Each boundary is a suture zone of earlier historical

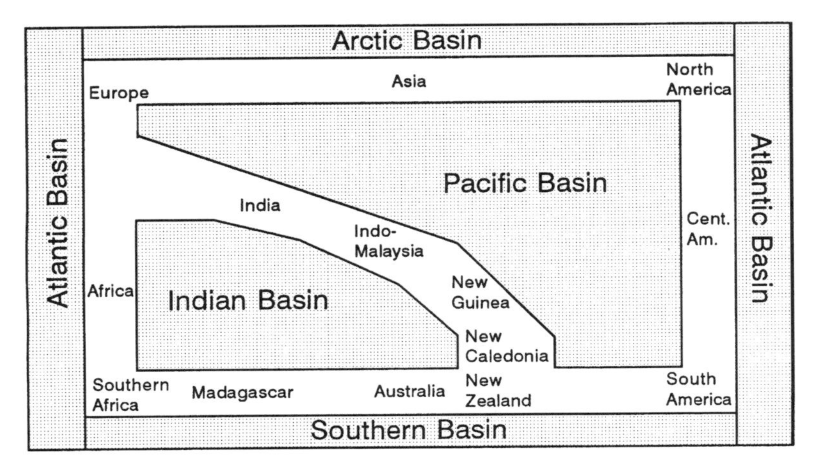

FIGURE 6-13. Formalized panbiogeographic classification of major biogeographic regions as ocean basins. Wallacean regions are now recognized as artificial conglomerates of geology and biology located at the boundaries of tectonic basins that provide the basis for a natural concept of biogeographic homology (from Craw and Page 1988: fig. 12; reprinted with permission of John Wiley & Sons).

landscapes (e.g., Parenti 1991). Despite cladistic interpretation of panbiogeographic areas as discrete units (Nelson 1985), the panbiogeographic system irreparably fractures the traditional biogeographic regions along new lines of demarcation.

The cartographic order of Wallace's zoogeographic and de Candolle's phytogeographic regions is replaced by the instability of the panbiogeographic border, a map of great transoceanic and intercontinental border zones in which no centers remain. The old categories of clearly defined geographic regions and neatly nested areas of endemism are faulty according to panbiogeography. A biogeographic region is in effect all boundary; boundaries pass everywhere, through every recognized region and area of endemism. As with more localized biogeographic nodes, which are both centers and boundaries of distribution, regions are also margins, and difference, convergence and hybridity underpin this conception and description of bioregions in flux.

7

Tracks, Nodes, Biodiversity, and Conservation

As many as half of all terrestrial species may become extinct over the next 50 years. Prominent ecologists recognize that if this crisis is to be managed, the present research emphasis on detailed individual species studies will need to be replaced with methods capable of evaluating entire communities and ecosystems (e.g., Ehrlich 1992). New research methods are needed to evaluate ecosystems, represent biodiversity patterns and processes, and develop the capabilities for preserving this biodiversity (Alkin and Winfield 1993). These different objectives are often subsumed within a general concept of "global biodiversity"—meaning a pattern of biodiversity that has a global structure. To meet this challenge, Ulfstrand (1992) called for the development of a research program of "space science magnitude" to determine how and why biodiversity varies in space and through time, in order to evolve the tools for conserving and managing it. However, the scientific foundations for this program are already available in biogeographic concepts, and methods for analyzing the geography and history of biodiversity are already in use by biogeographers (Bowman 1994).

Biodiversity is most often portrayed as a nongeographic or ahistorical quantity that is understood through counting the number of species or taxa (Harper and Hawksworth 1994, May 1994) or through developing simple ratios of species richness (Hammond 1994). Platnick (1992) observed, however, that the biodiversity question is really a biogeographic one because it is a question of where on this planet limited financial and human resources should be applied. Different animals and plants occur in different places, and these differences pose an "agony of choice" in the setting of conservation priorities (Vane-Wright et al. 1991, Crozier 1992). If biodiversity was localized to a few points in space, there would not be a conservation problem, but in the real world different localities exhibit varying levels of biodiversity and representation. Therefore, biodiversity analysis requires a research and management program with the conceptual and methodological resources able to deal simultaneously with a multitude of points or localities.

7.1 Biodiversity as Biogeography

Establishing a scientific conservation value requires criteria by which distribution data can indicate natural patterns of biodiversity so that preservation priorities and comprehensive management strategies can be formulated. Biogeography meets the "space science" requirements of Ulfstrand because it provides methods to describe patterns of biodiversity and also to understand the global biological and geological processes involved (see Nelson and Ladiges 1990, Grehan 1992, Bowman 1994, Lourenço and Blanc 1994). In addition, biogeography is cost effective because global biodiversity information can be developed from available data without losing time and money on redundant exploratory research and inventory compilation (Grehan 1991b).

There is a growing awareness that biogeographic data and analysis underpins the description and measurement of biodiversity values for conservation planning and sustainable resource management. This is because knowledge of the location of the components of biodiversity is absolutely critical to effective network design for *in situ* conservation (Margules and Austin 1994, Humphries et al. 1995). Raven and Wilson (1992) proposed a 50-year data collection program to completely inventory all the global biota, but simply accumulating locality records of taxa in database systems may be of little practical value (Renner and Ricklefs 1994). Instead, new biogeographically oriented databases need to be constructed to allow analysis of the geographical distributions of organisms.

A number of phylogenetic and systematic approaches are being developed to compare taxonomic difference and representation for different areas or localities (e.g., Vane-Wright et al. 1991, Pressy et al. 1993, Faith 1994, Williams and Humphries 1994). These methods are biogeographic in that they evaluate taxonomic diversity over geography, but the units of spatial comparison are themselves assumed rather than analyzed—the units of homology are based on genealogical relationships as estimated from molecular and morphological characters, rather than acknowledging spatial characters, so that the biodiversity "elements" are still solely taxa.

Despite the appeal of phylogenetic methods to prioritize conservation choices, the requisite cladistic studies and descriptions of new species will not be completed within the few years left for reserve designation and management in many critical areas. These methods require that a significant proportion of the cladogram for the organisms under consideration be partially if not fully resolved. This is a most unlikely prospect in the foreseeable future for the vast majority of the world's biota.

Panbiogeographic concepts and methods are applicable to research problems of local and global biodiversity because they provide suitable criteria for documenting, mapping, and tracking the natural spatial characteristics of biodiversity. Through panbiogeographic maps the conservation scientist and manager can have access to regional and global contexts as well as a ready resource of geographic information that identifies the natural biogeographic units of biodiversity necessary to establish research and management priorities. Panbiogeography is an ideal tool for the advocates and designers of protected natural areas, because it emphasizes the importance of geographic distribution. Its applications in conservation biology can aid the recognition of the relative regional, national, and international significance of the habitats requiring protection.

7.2 Mapping and Tracking Biodiversity

Conservation and biodiversity programs such as the International Union for the Conservation of Nature (IUCN) have already established global biodiversity maps that represent modified versions of Wallace's (1876) classification system (Udvardy 1975, 1987). In this, as in other traditional biogeographic schemes, the natural spatial structure of biodiversity is constrained within the geopolitical boundaries of the earth's major landmasses. These biogeographic classifications represent biodiversity as comprised of collections (areas or regions) of objects (animals and plants). Each area constitutes a discrete environmental container allocated a place in a hierarchy of biogeographic regions, provinces, biomes, ecological areas, and districts. In these systems, the spatiality of biodiversity is subordinated to a sense of order imposed by an hierarchical structure wherein organisms are separated from their environments. Boundaries are drawn across landscapes, ignoring the complexities of biogeographical and geological history. This acceptance of division, separation, and isolation as natural features of landscapes can exert a profound influence on the allocation of financial and scientific resources to biodiversity management (Grehan 1990a, 1993).

These current conventions of biodiversity classification are problematic because they impose a hierarchical system separating biodiversity from itself—the animals and plants are isolated from their geographical

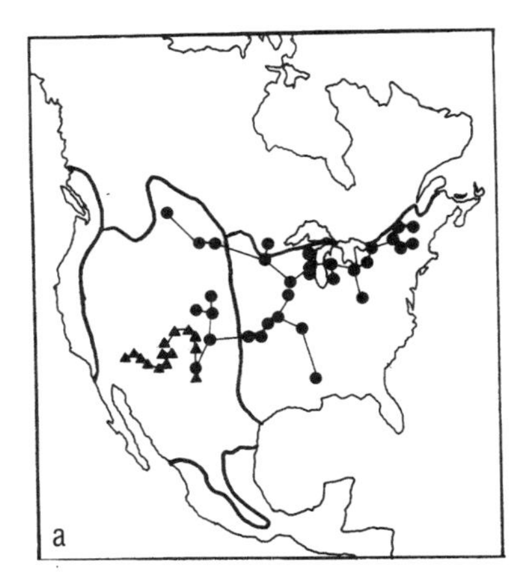

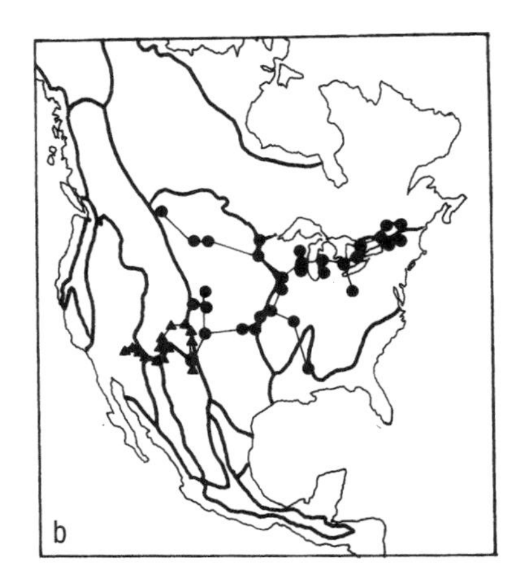

FIGURE 7-1. Tracks crossing the "biogeographic" boundaries (bold lines) in North America of (a) areas designated by Wallace (1876) and (b) the more complex International Union for the Conservation of Nature version. The tracks represent sister species of ground beetles (*Harpalus*: Carabidae) with a vicariant relationship (from Grehan 1995: fig. 3; reprinted with permission of Science and Management of Protected Areas Association).

and historical contexts that are represented only as enclosing containers. Such a classificatory structure implies a control and constraint upon the landscape that many conservation biologists would not support. These classifications impose boundaries that are arbitrary constructs that do not exist in the real, natural world. For example, tracks cross Wallacean and IUCN boundaries as if they did not exist as shown by a ground beetle (Carabidae) distribution in North America (fig. 7-1). The artificiality of these systems means that they cannot provide novel biodiversity insights, and neither are they relevant to developing global concerns (cf. Eldredge 1992).

Tracks, Nodes, and Biodiversity

A global panbiogeographic map represents biodiversity as a network of standard tracks and nodes (fig. 7-2). This approach to biodiversity classification can provide spatiotemporal units of biodiversity to mirror natural occurrence because distributions are classified according to unique spatiotemporal characters (see chapter 6). This concept of biogeographical homology introduces a new perspective for biodiversity studies where the spatiotemporal elements of biodiversity are the tracks, nodes, and baselines that identify different biogeographic centers and sectors. For global biodiversity this means that major units of

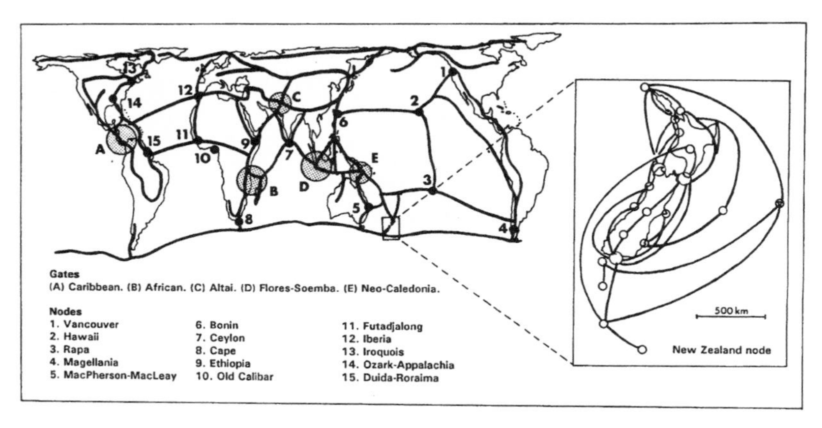

FIGURE 7-2. The geography of global biodiversity represented as a network of tracks and nodes (modified from Croizat 1958). Primary centers of biogeographic diversity are indicated by large circles (gates) and letters A–E. Other regional diversity centers are numbered 1–15. Inset: biodiversity centers and relationships summarized for the New Zealand archipelago (from Grehan 1995: fig. 7; reprinted with permission of Science and Management of Protected Area Association).

homology are the ocean basins, and it is from this classificatory base that the spatiality of local biodiversity may be studied in relation to earth history events that contribute to local differentiation and uniqueness. Miller (1994) identified tracks and nodes as an approach to mapping the diversity of nature that promotes understanding of the biosphere's evolutionary infrastructure as well as identifying biodiversity hot spots.

In panbiogeography the current concepts of biodiversity are being extended beyond being the additive property of elements (i.e., organisms) toward differences in spatial homology that divide and separate one type of distribution from another. Nodes are of particular importance because they are points of confluence where plants and animals may exhibit local presence such as endemism. In addition, they describe the phylogenetic and geographic limits of taxa and their relationships with other areas (Heads 1990). These properties of nodes correspond to the principal elements that define centers of biodiversity. Nodes are, therefore, biodiversity hot spots considered in a biogeographic context.

Five major nodes have been designated as "gates" to highlight their pivotal role in the evolution of current distributions, but many other biogeographic centers of diversity with global or regional significance are scattered over the earth (fig. 7-2). In conservation biogeography each gate could provide a focus for documenting and assessing local and regional diversity patterns. New World biodiversity, for example, is centered on the Caribbean gate, which orients tracks north and south, east and west across the Atlantic to Europe and Africa, and across the Pacific to East Asia and Australia. This track/node network or internet implies that ecological and geological events associated with the history of the Caribbean may have had a significant influence on ancestral distributions for much of the biota throughout North and South America. The biodiversity significance of local areas can only be adequately assessed by addressing the biotic patterns both inside and outside the designated area of concern. Comprehensive conservation of Amazonian biodiversity, for example, also requires a conservation policy for the Caribbean.

The Biodiversity Atlas Project

Panbiogeographic and biodiversity atlases have been proposed for New Zealand (Grehan 1990a) and southern South America (Morrone and Crisci 1992). Preliminary steps toward such a project have been implemented by Heads (1993, 1994a,b,c) for more than 160 species of New Zealand angiosperm plants. Documentation and analysis of tracks and nodes are appropriate for biodiversity studies if taxonomic and distribution data are available. An atlas could be compiled from distributions gleaned from taxonomic monographs and other resources such as reference collections (an approach successfully applied in Hawaii, for example; Mlot 1995). Records that are site or point specific (i.e., drawn as dot matrices or listed as geographic map coordinates) are most suitable because they allow the maximum level of geographic resolution for track and node analysis. Scientific collections held by private individuals could provide additional sources of valuable information, particularly where specialist assistance is available (Roscoe 1990).

Locality records can be stored in an electronic database as a system of geographic coordinates (such as latitude/longitude), facilitating integration with existing geographic information systems (GIS) and conversion into tracks. New locality data can also be easily incorporated as they become available from targeted inventory surveys. The atlas would comprise an analysis of the standard tracks and nodes for the region in question and could illustrate individual tracks as well as summaries of track and node analyses. Conservation significance of the atlas resides in the formulation of a value index system for tracks and nodes. Using vertex analysis (see chapter 5), the number of tracks involved with standard tracks and nodes can be quantified and the consequent values used to provide biogeographic criteria for ranking habitats and localities of a region.

This atlas could contribute to these priorities in two different directions—local and regional. The geographic coordinates of an area could be entered into the atlas database and compared with the distributions correlated with standard nodes and tracks. This relationship would indicate the number of standard tracks and nodes located within the designated area as well as their biogeographic ranking. Taxa related to the tracks and nodes could also be represented, along with their relationship to the nearest standard tracks and nodes outside the designated area. This information is particularly important when there is no distribution information for the area under consideration, and it will demonstrate any need for field surveys to enable further evaluation. Thus, a comprehensive biodiversity inventory on any spatial scale will include not only a listing and ranking of taxa, but also a listing and ranking of tracks and nodes. At the regional level the documentation of spatial geometry using track and node methodology will complement ecological studies of target species and systematic analyses of the genealogical relationships between areas and taxa within particular biogeographic patterns. Differences in phylogenetic representation may provide important indicators of nodal characteristics like boundary limits, relictual centers, unique endemics, and the absence of phylogenetically important groups.

Global, regional, and local syntheses of tracks and nodes can help conservation research policy by alerting managers to those localities that are significant for their biological values. Regional biodiversity issues can be addressed in terms of nodes or localities that exist in a biogeographic series rather than being isolated by political or administrative boundaries. In this respect the panbiogeographic atlas is different from the approach used in gap analysis, where gaps in reserve representation are identified by simple geographic overlap of species within specific political or administrative areas (e.g., countries, states) (see Scott et al. 1991). Although gap analysis may identify areas of high species diversity, the geographic patterns themselves are not subject to biogeographic analysis. Gap analysis also maps environmental characters (often vegetation) associated with particular animal species to predict potential new localities, although this approach is not always successful (Scott et al. 1993, Conroy and Noon 1996). Distribution gaps (whether sampling error or extinction) are displayed by the panbiogeographic atlas as tracks linking disjunct localities, and the geometry of a track also predicts the geographic range of new localities (i.e., gaps within the range of the track). The use of GIS techniques to overlay protected areas with known or predicted species ranges can be used in both gap analysis and the panbiogeographic atlas, but only in the latter is the local and global significance of species distributions directly accessible through analysis of biogeographic patterns.

7.3 Tracks and Nodes in Conservation Biology

Case Study 1: Nodes, Tracks, and a New Zealand National Park

Tracks and nodes can be incorporated into local conservation planning strategies by documenting biogeographic patterns associated with a particular area of conservation interest. In New Zealand, for example, a Department of Conservation assessment of northwest Nelson for national park status (Anon 1993) referred to the "huge" amount of endemism, but the area also overlaps with a statistically significant number of track boundaries (Henderson 1990, 1991). Through their biogeographic status, northwest Nelson landscapes are identified as some of the preeminent biodiversity centers in the country. Even without an established biodiversity atlas for New Zealand, some of the principal biogeographic elements can be recognized using landsnail distributions (Grehan and Climo 1992).

Endemism is rarely as absolute as portrayed in conservation articles, and a seemingly isolated, internal pattern will have biogeographic representation elsewhere. For instance, two snail species "endemic" to the Kahurangi (northwest Nelson) National Park boundaries are represented outside the park by a related third species in the southern South Island (fig. 7-3a). Individual species may be endemic, but their track relationships are not.

Geopolitical limits of northwest South Island are entirely arbitrary, so even though a distribution may not be endemic to the area, its main massing there may represent a significant biogeographic element. The main massing of *Pseudegestula worleyi*, for example, is concentrated largely to the west of the Alpine fault (fig. 7-3b), a major tectonic feature that is significant for the biogeography of many New Zealand organisms (see chapter 2). This snail is also of phylogenetic significance as one of only three or four New Zealand species that broods embryos in the umbilical well on the shell base (Grehan and Climo 1992).

Numerous tracks connect northwest Nelson with Okarito and/or Northland along the western coastline of New Zealand. The two species of *Phrixgnathus* within the park boundaries represent a western and eastern track, respectively (fig. 7-3c). A third species extends this track structure to Northland. The parallel tracks illustrated for *Phrixgnathus* may be extended farther within the New Zealand archipelago. A group of three as yet undescribed genera of molluscs exhibit two parallel arcs that extend along the west side of the South Island, with the eastern arc represented by a single record at Mt. Arthur (an important biogeographic center within the park boundary) and to the north (Northland) and far south (subantarctic islands) (fig. 7-3d). In these examples northwest Nelson represents a key center of differentiation for taxa, even though none of its constituents are "endemic" to the park itself.

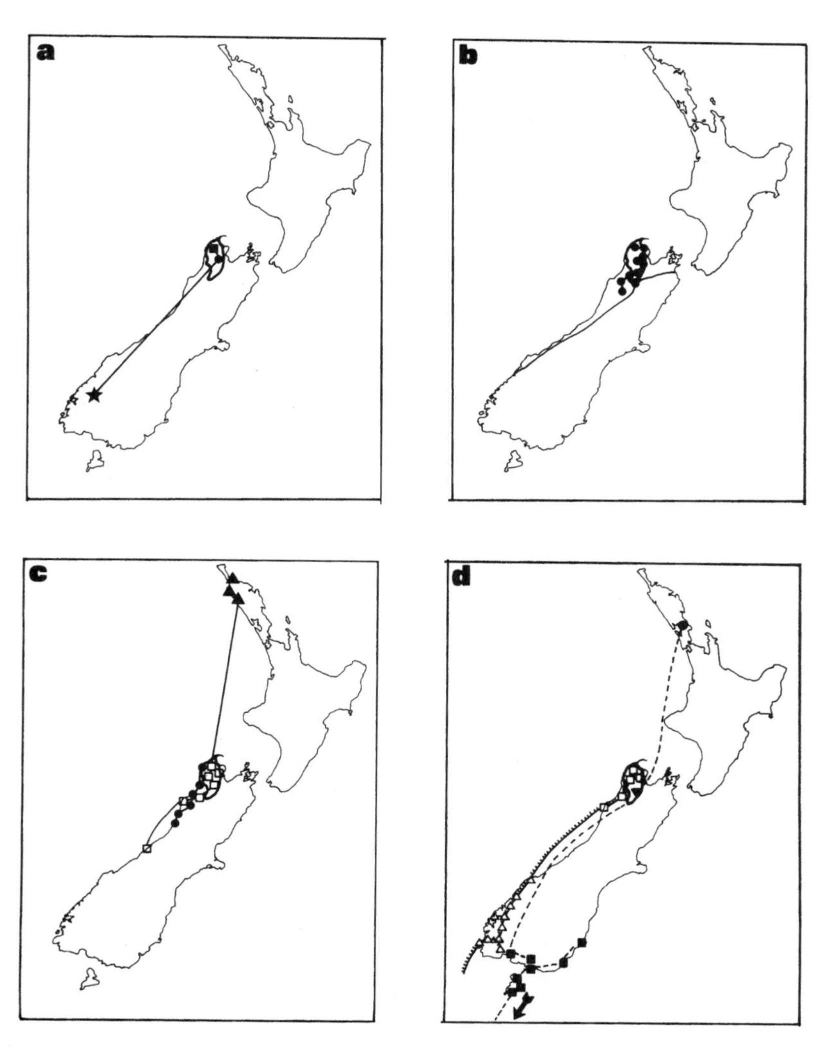

FIGURE 7-3. Biogeographic elements of northwest Nelson (enclosed by solid line), New Zealand, represented landsnail tracks. (a) Local cave-dwelling endemics, *Zelandiscus elevata* (Aorere Valley, square), Z. *n.sp.* (Upper Takaka, circle), Z. *worthyi* (Te Anau, star); (b) *Pseudegestula worleyi* (Charopidae), Alpine fault as solid line; (c) *Phrixgnathus marginata* group (Punctidae), tracks *P. marginata* (filled circles), P. *n.sp.* (open squares), and *P. larochei* (filled triangles); (d) parallel arcs for snail species in three related genera (all taxa undescribed): western arc (hatched line) for genus 1 (open squares, triangles); eastern arc (dashed line) for genus 2 (filled circle), and genus 3 (filled inverted triangle, filled squares). Arrow indicates extension of eastern track to subantarctic islands (not shown on map) (after Grehan and Climo 1992).

FIGURE 7-4. Biogeography of *Nothofagus* in New Zealand. (Left) Revolute-leaved group: *N. fusca* (hatched line), *N. solandri* (stippled line), *N. truncata* (cross-hatched areas). Numbered nodes and solid black areas (all three species present): 1, Arawata; 2, northwest South Island; 3, Wellington; 4, Bay of Plenty. (Right) Plane-leaved species: *N. menziesii* (after Heads 1990: fig. 17b; reprinted with permission of SIR Publishing).

Biogeographic tracks are ecological as well as historical, and this ecology may also be tracked. Southern beeches (*Nothofagus*: Nothofagaceae) comprise a dominant ecological element of the northwest Nelson forest landscape. Although widespread, with extensive overlapping distributions in New Zealand, the biogeography of this group also emphasizes the importance of northwest Nelson (Heads 1990). The three New Zealand species of *Nothofagus* sharing revolute-shaped leaves occur together only at four disconnected massings (fig. 7-4). Northwest Nelson is included within one massing and is adjacent to the Wellington massing across the Cook Strait center. The southern massing at the Arawata node is closely associated with the Alpine fault zone as is northwest Nelson, north of the Wairau fault. A single plane-leaved *Nothofagus* species is widely distributed in New Zealand from the Bay of Plenty massing southward, and it is conspicuously present south of the Arawata node in an area biogeographically related to where the other plane-leaved South American *Nothofagus* are found. These correlations between *Nothofagus* distribution and tectonic features reflect a mutual historical and ecological origin for the presence of *Nothofagus* in the northwest South Island that is precisely connected by standard tracks to the ecology and history of New Zealand in general.

Many more biogeographic patterns for a diverse assemblage of animal and plant taxa conform to a similar spatial geometry (e.g., Craw 1989, Gibbs 1990, Heads 1990) and emphasize the key biogeographic position of northwest Nelson. Northwest Nelson is a major biogeographic gate, being involved with events that have shaped biodiversity both within New Zealand and beyond from Mesozoic times to the present. These patterns and insights may be recognized by conservation managers through understanding the spatial patterns as a real constituent of biodiversity.

Case Study 2: Tracks and Nodes in Conservation Practice

Butterflies are an attractive, conspicuous, and taxonomically well known group of organisms for which good distribution information exists. They are good indicators of environmental health as well as being sensitive to climate and hence are important bioindicators of climatic change (Dennis 1995, Pullin 1995). Internationally butterflies have featured as invertebrate "flagships" for conservation as well as ecotourism. Although at present no endemic butterflies are causes for major conservation concern in New Zealand, some species may be considered at long-term risk from habitat alteration, introduced parasitoid wasps, and global climatic change (New 1995). Identifying sites for butterfly reserves and population monitoring will be an essential part of any comprehensive policy for the preservation of New Zealand's biodiversity.

Maximal generic and species diversity of endemic New Zealand butterflies is found in the South Island (Gibbs 1980). Track analyses of the species with the most restricted distributions in the South Island identifies the Spenser Ecological District as nodal for these butterflies (fig. 7-5). This area contains the southern-most distributional limits for two species, the forest ringlet (*Dodonidia helmsii*) and Harris's tussock butterfly (*Argyrophenga harrisi*), and the northern limits for Butler's ringlet (*Erebiola butleri*). Two other species, a tussock butterfly (*Argyrophenga antipodum*) and a copper (*Lycaena rauparaha*) approach this node along an arc extending from the northeast to the southeast of the Spenser district. In addition, another tussock butterfly (*A. janitae*), the black mountain butterfly (*Percnodaimon merula*), the red admiral (*Bassaris gonerilla*), and three coppers (*L. boldenarum, L. feredayi, L. salustius*) occur within this district's boundaries and are also widespread elsewhere. No other region of New Zealand has such a diversity of the nation's butterfly genera, and it is clearly the major biogeographic hot spot for butterflies in a national context. Butterfly reserve design and establishment, further inventory, monitoring and survey to assess long-term indicator trends in

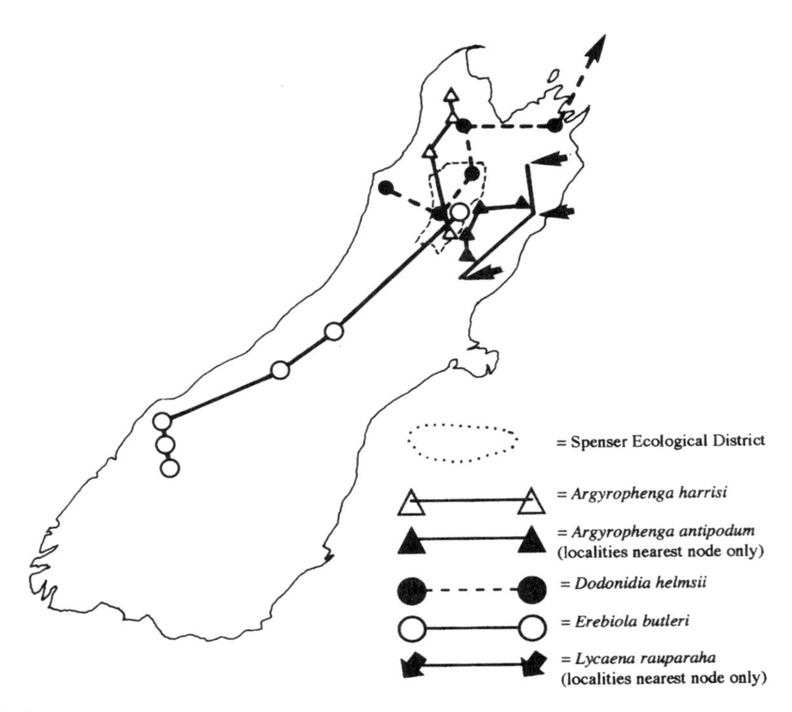

FIGURE 7-5. Location of primary node for endemic New Zealand butterfly genera in the Spenser Ecological District (indicated by dashed line), South Island. Tracks for taxa with restricted distributions represented by *Argyrophenga harrisi* (open triangles and solid line), *Dodonidia helmsii* (filled circles and broken line; arrow by track indicates extension to North Island), *Erebiola butleri* (open circles and solid line), *Lycaena rauparaha* (arrows and solid line, localities nearest node only) (data from Gibbs 1980, Craw 1978b).

relation to climate and other environmental conditions would be most cost effective if focused on this node and environs. Heads (1994a) used track and node analysis to demonstrate that the Spenser district possesses maximum diversity for the Australasian angiosperm shrub genus *Parahebe*, thus indicating its potential national and regional significance for general biodiversity and conservation studies.

Case Study 3: Tracks and Nodes in Conservation Management

Situated in the southeast of New Zealand's South Island are the Lammermoor/Lammerlaw Ranges, an area of uplifted Cretaceous peneplain characterized by endemic species, high species and generic diversity, and the absence of taxa that are widespread elsewhere in southern montane

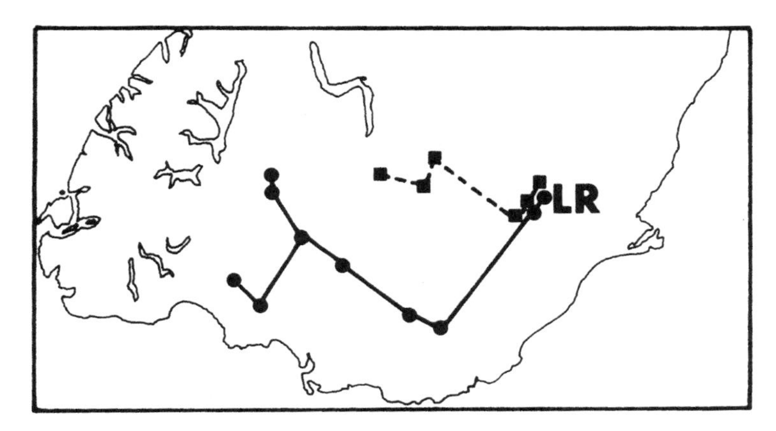

FIGURE 7-6. Nodal status of Lammermoor Ranges (LR) demonstrated by track overlap of two sister species of crambine moths, *Orocrambus geminus* (squares and dashed line) and *O. scutatus* (circles and solid line) (Lepidoptera: Crambidae) (from Patrick 1991: fig. 4; reprinted with permission of SIR Publishing).

regions (Barratt and Patrick 1987, Patrick et al. 1993). These flat-topped mountains have been identified as a biogeographic node of major significance for numerous insect groups and several plant groups. They constitute one of the major centers of diversity for a genus of crambine moths (*Orocrambus*), posses a diversity of diurnal geometrid moths unparalleled elsewhere in New Zealand, and are of nodal significance for the angiosperm shrub genus *Leonohebe*. The thymelaeaceous plant genus *Kelleria*, with an Indo-Australian distribution from Borneo to subantarctic Auckland Islands is represented by six species—the most at any locality within the entire generic range (Heads 1991). Using track analysis, Patrick (1991) confirmed the nodal status of these ranges (fig. 7-6), after recommending that they be designated a national reserve (Patrick 1984).

On a more local scale, track analysis has been used to assess the conservation values for invertebrates of Matiu and Mokopuna Islands in New Zealand's Wellington Harbour. Grehan (1990c) showed that although the insect biota of these islands was neither unique nor endangered, it was of intrinsic biogeographic and scientific interest and of potential conservation management significance. Biogeographic tracks involving the southern half of the North Island and the eastern half of the South Island were demonstrated for several species of beetle and a hemipteran (fig. 4-7). This southern North Island–South Island generalized track is known for at least 60 other animal and plant species (Craw 1988). These distributional tracks connect coastal and insular localities with inland or upland habitats, resulting from major processes of tectonic uplift with associated alterations in ecological amplitude for in-

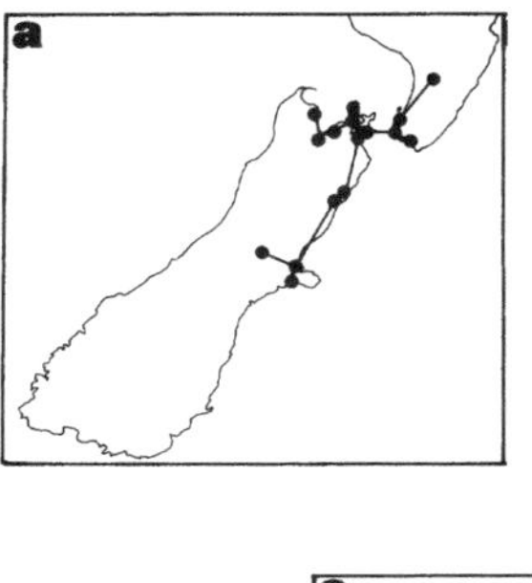

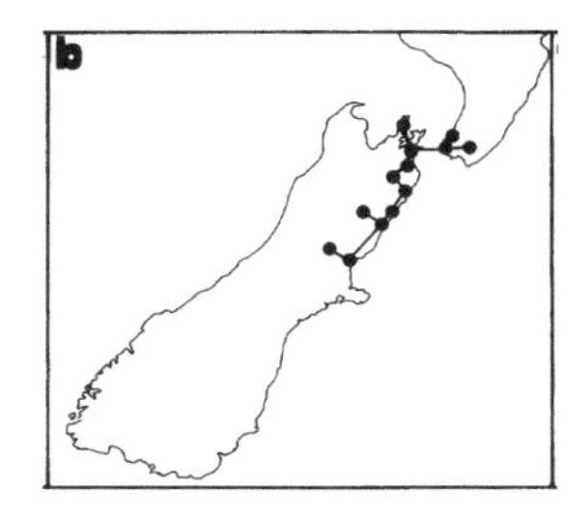

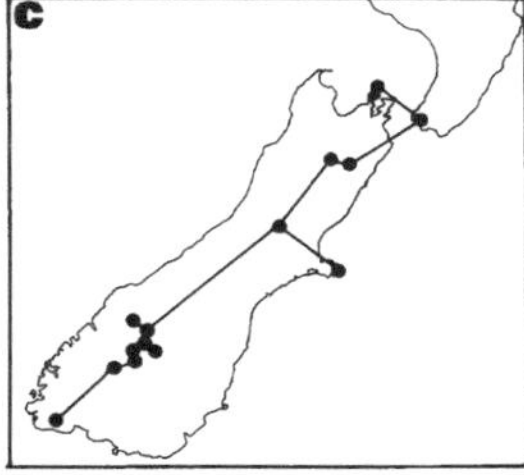

FIGURE 7-7. Tracks for some Matiu (Somes) and Mokopuna island insects. (a) *Dysnocryptus pallidus* (Coleoptera: Anthribidae), (b) *Mecodema sulcatum* (Coleoptera: Carabidae), (c) *Hudsona anceps* (Hemiptera) (after Grehan 1990b: fig. 6; reprinted with permission of *New Zealand Entomologist*).

cluded taxa. Wellington populations of component taxa of these tracks, other than those on the islands, are now fragmented and scattered throughout the highly developed Wellington region, which has been a site of major landscape alteration through human activity. Knowledge of these biogeographic relationships is of major importance for determining what animal and plant species should be translocated from nearby sites as part of any ecological restoration project on the islands (Grehan 1990b).

7.4 Hybrids, Species, and Conservation

Historical biogeographical issues directly impact on policies and procedures affecting the conservation of species and other taxa. Species have traditionally been defined by spatiotemporally independent criteria such as mate recognition, reproductive isolation, and unique genetic or morphological characters. Many philosophers of biology and systematics now recognize species as phylogenetic individuals with particular boundaries in space and time; they can be diagnosed with respect to particular places and times, but not defined by non-spatiotemporal transcendental criteria (Ghiselin 1974, Hull 1976, Ereshefsky 1992).

This understanding of species as phylogenetic individuals constitutes a major challenge to traditional approaches to species conservation and to the management response to interspecies contact and breeding. Conservationists have traditionally managed interspecies hybridization in two ways. Either the species are not "true" species after all (i.e., erroneous taxonomy), or, if the species are true, hybridization should be prevented (O'Brien and Mayr 1991).

The critical nature of species definition for conservation is forcefully illustrated by the case of the New Zealand "tobin," a Chatham Island black robin–Chatham Island tit hybrid hatched in January 1990. The female parent was a fosterling given to Chatham Island tits to be reared as part of a recovery program for the black robin population, one of the world's most endangered bird species. This female fosterling confounded expectations by mating with a tit instead of another black robin. The tobin was duly photographed and then killed—a procedure justified by declaring the hybrid a threat to black robin survival: "Unfortunately, Tobin would have to be destroyed, for the team could not risk her being fertile and surviving to breed, say, with a robin. New Zealand had enough problems in the past with two of its endangered birds, the black stilt and Forbes' parakeet, hybridizing with near relatives" (Butler and Merton 1992:235).

This justification presumes that each species exists in an ideal state that must be preserved. The notion that a hybridizing species is undesirable is responsible for hybridizing species being unprotected under the U.S. Endangered Species Act (Grant and Grant 1992). But hybridization is not so easily dismissed, as interspecies hybridization occurs in at least 10% of all bird species (Grant and Grant 1992), and hybridization has been recognized as a significant speciation mechanism in both invertebrates and vertebrates (Bullini 1994).

Elimination of the tobin was justified by labeling it as a "freak event" that would not normally occur in nature—this event being the fact of human intervention in transferring black robin eggs into tit nests. Humans are seen to constitute a foreign element in the landscape, but such an outlook is biogeographically problematic because no natural landscapes have escaped the impact of humans (Newbigin 1912, Cockburn 1989). There is no biogeographic imperative requiring conservationists to impose boundaries between artificial and natural landscapes or situations. Instead there is a biogeographic interest in focusing on the integrity of tracks and nodes that are bound up with the structural and functional character of existing ecosystems and associated organisms.

Over the natural range of many species, different populations may interbreed in one set of localities but not in another. No characters, whether biochemical, morphological, ecological, geographical, or genetic, are both necessary and sufficient to define a species, as species are variable and varying (Gray 1988). Whether or not humans are directly involved, species may hybridize under one set of circumstances and not in another. The biogeographic reality of evolution is that polytypic speciation through vicariant form-making implies that all "species" are biogeographic and phylogenetic composites, and hybridization is simply another aspect of ancestral and parental expression that has no proper name or formal identity (Craw 1993, Craw and Hubbard 1993). In this context, conservation scientists may consider hybridization as a possibly desirable outcome for endangered species. The tobin hybrid could have promoted survival of the black robin lineage through increased genetic heterozygosity and evolutionary novelty. Mallet (1995) suggests that the huge amount of genetic biodiversity found in infraspecific taxa should not be ignored and that the conservation of actual morphological, ecological, and genetic diversity should take precedence over a nebulous taxonomic category about which there has been so much disagreement. Moritz (1995) concluded from a molecular perspective that translocation of individuals should focus on maintaining overall processes, such as historical levels of gene flow in the context of historical hybridization, rather than on specific entities.

7.5 Conclusions

The proposed panbiogeographic biodiversity atlas is scientifically defensible in the sense that it meets the criteria of objectivity, repeatability, and testability. It would also be useful in a practical context because it addresses two of the most urgent concerns in conservation biology through documentation of biodiversity patterns and making the significance of those patterns available in a form that can be used by conservation managers. General advantages of the panbiogeographic biodiversity atlas can be summarized as follows:

1. The biodiversity atlas could contribute standard criteria for describing the spatial structure of biodiversity in the form of track and node maps and matrices and thus facilitate scientifically defensible conservation and resource management decisions,
2. Applications of the methods will enable conservation choices and priorities to be made through quantification of track and nodal values as natural measures of biodiversity in space and time,
3. These methods are predictive about the evolution of biotas, as has been demonstrated by their generation of novel predictions about historical patterns and processes responsible for current distribution (Craw and Weston 1984).

Development of management policies that place local areas in regional perspective using track and node networks requires the design and implementation of a biodiversity atlas that has no natural limits other than on a global scale. The representation of natural biodiversity patterns may have to start at the global level and work down to local and regional concerns through the representation of tracks and nodes. A refined version of the global biogeodiversity map (fig. 7-2) could provide an effective foundation for a biodiversity program generating rapid and significant results at a low cost compared with inventory programs such as those proposed by Raven and Wilson (1992). Inventory is a necessary and continuing component of biodiversity studies, but it may be most effective if targeted to specific problem localities identified by track and node analyses. Additional data can then be effectively incorporated into an analytical biodiversity program. Biodiversity is made up not of genes, species, and habitats, but of the tracks and nodes of life.

Glossary of Technical Terms

antinode a node or distribution area characterized by the absence of taxa found widely elsewhere.

baseline a geographical or geological feature of an individual track such as crossing an ocean or sea basin, or a major tectonic feature (e.g., fault zone), that is interpreted as a diagnostic character (i.e., spatial homology) uniting individual tracks that may otherwise have little in common (e.g., no distributional congruence).

clade a monophyletic lineage or group.

cladogram a branching diagram relating terminal taxa based on the distribution of derived character states among the taxa.

clique (1) a set of objects in which each individual object is related in some way to every other object; (2) a set of vertices in a graph of objects, each of which is adjacent (i.e., connected to each other).

compatible (1) specifying the same set of relationships; (2) adjacent in a graph.

concerted evolution nonindependent evolution of repetitive DNA sequences resulting in a sequence similarity of repeating units that is greater within than among species.

gate major biogeographic node at the intersection of two or more standard transoceanic tracks.

generalized track *see* standard track.

grid analysis a quantitative method for calculating the geographic location of main massings.

immobilism a condition in which the distribution boundaries are in a stable state.

incompatible (1) specifying different sets of relationships; (2) not adjacent in a graph.

laws of growth an evolutionary process proposed by Darwin in which structures and adaptations owe their origin to principles of biological organization rather than to natural selection. Equivalent to modern concepts of developmental, genetic, and phylogenetic constraints.

main massing a locality or distribution area possessing the greatest concentration of diversity within the geographic range of a taxon.

minimal spanning tree an acyclic graph that connects all the localities or distribution areas occupied by a taxon or characters such that the sum of the length of the links connecting all localities or distribution areas is the smallest possible.

mobilism a condition in which the distribution boundaries of a taxon are in a state of flux.

molecular drive DNA turnover dynamics that promote improbable long-term shifts in the mean genotype of a population.

nodal value the sum of the number of links a particular distribution area or locality has with other distribution areas or collection localities in a track graph.

node (1) an area or locality where two or more standard tracks overlap; (2) an area or locality in a track graph with an higher than average connectivity value.

parsimony a methodological principle that observed data should be explained in the simplest possible way (otherwise known as "Ockham's razor").

spatial homology *see* baseline.

standard track a set of two or more individual tracks that are compatible or congruent according to a specified criterion (e.g., shared baselines or compatible track geometries).

Steiner tree an acyclic graph connecting a set of points in which additional points are inserted to minimize the length. An example is a cladogram of taxa in which the additional points can represent hypothetical ancestors.

track a line graph drawn on a map connecting the localities or distribution areas of a particular taxon or group of taxa.

vertex a point in a graph.

vertex analysis a quantitative method for calculating nodal values.

wing dispersal biogeographic pattern in which related taxa at opposite ends of the periphery of a distribution or track share a greater similarity with each other than they do to closer or adjacent related taxa.

References

Ackery, P. R. and Vane-Wright, R. I. (1984). *Milkweed butterflies: their cladistics and biology*. Cornell University Press, Ithaca, New York.

Airy Shaw, H. K. (1963). Notes on Malaysian and other Asiatic Euphorbiacae. *Kew Bulletin* 16, 41–372.

Allen, J. A. (1871). On the mammals and winter birds of East Florida. *Bulletin of the Museum of Comparative Zoology* 2, 161–450.

Allen, R. T. (1980). A review of the subtribe Myodi: description of a new genus and species, phylogenetic relationships, and biogeography (Coleoptera, Carabidae: Pterostichini). *Coleopterists' Bulletin* 34, 1–29.

Allen, R. T. (1990). Insect endemism in the interior highlands of North America. *Florida Entomologist* 73, 539–569.

Allkin, B., and Winfield, P. (1993). Cataloging biodiversity: new approaches to old problems. *Biologist* 40, 179–183.

Alvarenga, H. (1983). Uma ave ratite do Paleoceno Brasileiro. *Boletim do Museu Nacional de Rio de Janeiro Geologia* 41, 1–7.

Amorin, D. S. (1991). Refuge model simulations: testing the theory. *Rivista Brasilia Entomologia* 35, 803–812.

Anderson, N. M. (1982). The semiaquatic bugs (Hemiptera: Gerromorpha). Phylogeny, adaptations, biogeography, and classification. *Entomonograph* 3, 1–455.

Anderson, N. M. (1989). The coral bugs, genus *Halovelia* Bergroth (Hemiptera, Veliidae). II. Taxonomy of the *Himalaya* group, cladistics, ecology, biology and biogeography. *Entomologica Scandinavica* 20, 179–227.

Anon. (1993). Northwest South Island national park investigation. Report to the New Zealand Conservation Authority. Nelson/Marlborough Conservancy Management Planning Series No. 5, Nelson.

Askevold, I. S. (1991). Classification, reconstructed phylogeny, and geographic history of the New World members of *Plateumaris* Thomson, 1859 (Coleoptera: Chrysomelidae: Donaciinae). *Memoirs of the Entomological Society of Canada* 157.

Armstrong, P. (1983). The disjunct distribution of the genus *Adansonia* L. *National Geographic Journal of India* 29, 141–163.

Asquith, A. (1993). Patterns of speciation in the genus *Lopidea* (Heteroptera: Miridae: Orthotylinae). *Systematic Entomology* 18, 169–180.

Asquith, A., and Lattin, J. D. (1990). *Nabiula* (*Limnobaris*) *propinqua* (Reuter) (Heteroptera: Nabidae): dimorphism, phylogenetic relationships and biogeography. *Tijdschrift voor Entomologie* 133, 3–16.

Aubréville, A. (1936). *La flore forestière de la Côte d'Ivoire*, 3 vols. Larose, Paris.

Aubréville, A. (1969). Essais sur la distribution et l'histoire des angiospermes tropicales dans le monde. *Adansonia, Series 2*, 9, 189–247.

Aubréville, A. (1971). Essai sur la géophylétique de Manilkarées. *Adansonia, Series 2*, 2, 251–265.

Aubréville, A. (1974a). Les origines des angiospermes. *Adansonia, Series 2*, 14, 5–27.

Aubréville, A. (1974b). Origines polytopiques des angiospermes tropicales 2. *Adansonia, Series 2*, 14, 1455–198.

Avise, J. C. (1992). Molecular population structure and the biogeographic history of a regional fauna: a case study with lessons for conservation biology. *Oikos* 63, 62–76.

Avise, J. C. (1994). *Molecular markers, natural history and evolution*. Chapman and Hall, New York.

Avise, J. C., Arnold, J., Ball, R. M. Jr, Bermingham, E., Lamb, T., Neigel, J. E., Reeb, C. A., and Saunders, N. C. (1987). Intraspecific phylogeography: the mitochondrial DNA bridge between population genetics and systematics. *Annual Review of Ecology and Systematics* 18, 489–522.

Axelrod, D. I. (1972). Ocean-floor spreading in relation to ecosystematic problems. *University of Arkansas Museum Occasional Paper* 4, 15–68.

Axelrod, D. I. (1986). Cenozoic history of some western American pines. *Annals of the Missouri Botanical Gardens* 73, 565–641.

Axelrod, D. I., and Raven, P. (1978). Late Cretaceous and Tertiary vegetation history of Africa. In *Biogeography and Ecology of Southern Africa* (ed. M. Werger), pp. 77–130. Junk, The Hague.

Backus, R. H. (1985). Biogeographic boundaries in the open ocean. *Unesco Technical Papers in Marine Science* 49, 9–13.

Bae, Y. J., and McCafferty, W. P. (1991). Phylogenetic systematics of the Potamanthidae (Ephemeroptera). *Transactions of the American Entomological Society* 117, 1–143.

Bailey, T. C., and Gatrell, A. C. (1995). *Interactive spatial data analysis*. Longman Scientific and Technical, Harlow, Essex, England.

Ball, I. R. (1976). Nature and formulation of biogeographical hypotheses. *Systematic Zoology* 24, 407–430.

Banarescu, I. P. (1990). *Zoogeography of freshwaters, volume 1. General distribution and dispersal of freshwater animals*. AULA-Verlag, Wiesbaden.

Bandoni, S. M., and Brooks, D. R. (1987). Revision and phylogenetic analysis of the Amphilinidea Poche, 1922 (Platyhelminthes: Cercomerica: Cercomecomorpha). *Canadian Journal of Zoology* 65, 1110–1128.

Bardack, D. (1991). First fossil hagfish (Myxinoidea): a record from the Pennsylvanian of Illinois. *Science* 254, 701–703.

Barendse, W. (1984). Speciation in the genus *Crinia* (Anura: Myobatrachidae) in southern Australia: Phylogenetic analysis of allozyme data supporting endemic speciation in southwestern Australia. *Evolution* 38, 1238–1250.

Barratt, B. I. P., and Patrick, B. H. (1987). Insects of snow tussock grassland on the East Otago Plateau. *New Zealand Entomologist* 10, 69–98.

Battin, T. (1992). Geographic variation analysis amongst populations: the case of *Platycnemis pennipes* (Pallas 1771) (Insecta: Odonata: Zygoptera) in the Aegean. *Journal of Biogeography* 19, 391–400.

Bauer, A. (1990). Phylogeny and biogeography of the geckos of southern Africa and the islands of the western Indian Ocean: a preliminary analysis. In *Vertebrates in the tropics* (eds. G. Peters and R. Hutterer), pp. 275–284. Museum Alexander Koenig, Bonn.

Baum, D. A. (1995). A systematic revision of Adansonia (Bombacaceae). *Annals of the Missouri Botanical Garden* 82, 440–470.

Baumel, F. (1989). Description de *Rhithrodytes* nouveau genre d'Hydroporinae d'Europe et d'Afrique de Nord: Analyse phylogénétique et biogéographie (Coleoptera: Dytiscidae). *Annales Societe Entomologique de France* (n.s.) 25, 481–503.

Beard, J. S. (1953). The savannah vegetation of northern tropical America. *Ecological Monographs* 33, 149–215.

Belasky, P., and Runnegar, B. (1994). Permian longitudes of Wrangellia, Stikinia, and Eastern Klamath terranes based on coral biogeography. *Geology* 22, 1095–1098.

Benton, M. J. (1993). *The fossil record 2*. Chapman and Hall, London.

Bews, J. W. (1921). Some general principles of plant distribution as illustrated by the South African flora. *Annals of Botany* 35, 1–36.

Blake, D. B., and Zinmeister, W. J. (1988). Eocene asteroids (Echinodermata) from Seymour Island, Antarctic peninsula. *Geological Society of America, Memoirs* 169, 489–498.

Bledsoe, A. H. (1988). A phylogenetic analysis of postcranial skeletal characters of the ratite birds. *Annals of Carnegie Museum* 57, 73–90.

Blondel, J. (1988). L'oiseau, l'espace et le temps en Méditerranée. *Alauda* 56, 401–402.

Bond, W. J., Yeaton, R., and Stock, W. D. (1991). Myrmecochory in Cape fynbos. In *Ant-Plant Interactions* (eds. C. R. Huxley and D. F. Cutler), pp. 448–462. Oxford University Press, Oxford.

Borgmeir, T. (1957). Basic questions of systematics. *Systematic Zoology* 6, 53–69.

Borkin, L. J. (1986). Pleistocene glaciations and western-eastern Palearctic disjunctions in amphibian distribution. In *Studies in herpetology* (ed. Z. Rocek), pp. 63–66. Charles University Press, Prague.

Boucot, A. J. (1990). *Evolutionary paleobiology of behaviour and coevolution*. Elsevier, Amsterdam.

Boughey, A. S. (1957). *The origin of the African flora*. Oxford University Press, London.

Bowman, D. (1994). Cry shame on all humanity. *New Scientist* 1952, 59.

Bramwell, D. (1990). Panbiogeography of the Canary Islands flora. *Atti dei Convegni Lincei* 85, 157–166.

Brenan, J. P. M. (1978). Some aspects of the phytogeography of tropical Africa. *Annals of the Missouri Botanical Garden* 65, 437–478.

Breteler, F. (1993). *Novitates Gabonensis* 16: *Dichopetalum rabense* new species Dichopetalaceae from Gabon. *Bulletin du Jardin Botanique National de Belgique* 62, 191–195.

Briggs, J. C. (1987). *Biogeography and plate tectonics*. Elsevier, Amsterdam and New York.

Briggs, J. C. (1991). Historical biogeography: the pedagogical problem. *Journal of Biogeography* 18, 3–6.

Brignoli, P. M. (1983). Dispersion, dispersal, and spiders. *Verhandlungen des naturwissenshaftlichen veriens in Hamburg (NF)* 26, 181–186.

Britten, H. B., and Brussard, P. F. (1992). Genetic divergence and the Pleistocene history of the alpine butterflies *Boloria improba* (Nymphalidae) and the endangered *Boloria acrocnema* (Nymphalidae) in western North America. *Canadian Journal of Zoology* 70, 539–548.

Brooks, D. R. (1977). Evolutionary history of some plagiorchioid trematodes of Anurans. *Systematic Zoology* 26, 277–289.

Brooks, D. R. (1979). Testing hypotheses of evolutionary relationships among parasites: the digeneans of crocodilians. *American Zoologist* 19, 1225–1238.

Brooks, D. R. (1981). Fresh-water stingrays (Potamotrygonidae) and their helminth parasites: testing hypotheses of evolution and coevolution In *Advances in cladistics* (eds. V. A. Funk and D. R. Brooks), pp. 147–175. The New York Botanical Garden, Bronx, New York.

Brooks, D. R. (1985). Historical ecology: a new approach to studying the evolution of ecological associations. *Annals of the Missouri Botanical Garden* 72, 660–680.

Brooks, D. R. (1988). Scaling effects in historical biogeography: a new view of space, time, and form. *Systematic Zoology* 37, 237–244.

Brooks, D. R., and McLennan, D. A. (1991). *Phylogeny, ecology and behavior: a research program in comparative biology*. University of Chicago Press, Chicago.

Brower, A. V. Z. (1994). Rapid morphological radiation and convergence among races of the butterfly *Heliconius erato* inferred from patterns of mitochondrial DNA evolution. *Proceedings of the National Academy of Sciences* 91, 6491–6495.

Brown, J. H. (1986). Two decades of interaction between the MacArthur-Wilson model and the complexities of mammalian distributions. *Biological Journal of the Linnean Society* 28, 231–251.

Brown, J. H. (1995). *Macroecology*. University of Chicago Press, Chicago.

Brown, J. H., and Maurer, B. A. (1989). Macroecology: the division of food and space among species on continents. *Science* 243, 1145–1150.

Brown, J. W., Donahue, J. P., and Miller, S. E. (1991). Two new species of geometrid moths (Lepidoptera: Geometridae: Ennominae) from Cocos Island, Costa Rica. *Natural History Museum of Los Angeles County Contributions in Science* 423, 11–18.

Brummitt, R. K. (1968). A new genus of the tribe Sophoreae (Leguminosae) from western Africa and Borneo. *Kew Bulletin* 22, 375–386.

Brundin, L. (1981). Croizat's panbiogeography versus phylogenetic biogeography. In *Vicariance biogeography: a critique* (eds. G. Nelson and D. E. Rosen), pp. 94–158. Columbia University Press, New York.

Bullini, L. (1994). Origin and evolution of animal hybrid species. *Trends in Ecology and Evolution* 9, 422–426.

Burrows, J. E. (1990). *Southern African ferns and fern allies*. Sandton: Frandsen.

Burton, J. A. (1973). *Owls of the World*. P. Lowe Publications.

Bush, M. B. (1994). Amazonian speciation: a necessarily complex model. *Journal of Biogeography* 21, 5–17.

Butler, D., and Merton, D. (1992). *The black robin: saving the world's most endangered bird*. Oxford University Press, Auckland.

Cadle, J. E., and Greene, H. W. (1993). Phylogenetic patterns, biogeography, and the ecological structure of neotropical snake assemblages. In *Species diversity in ecological communities: historical and geographical perspectives* (eds. R. E. Ricklefs and D. Schluter), pp. 281–293. The University of Chicago Press, Chicago.

Cain, S. A. (1943). Criteria for the indication of center of origin in plant geographical studies. *Torreya* 43, 132–154.

Cain, S. A. (1944). *Foundations of plant geography*. Harper and Row, New York.

Camp, W. H. (1948). Distributional patterns in modern plants and the problems of ancient dispersals. *Ecological Monographs* 17, 159–183.

Cannatella, D. C. (1980). A review of the *Phyllomedusa buckleyi* group (Anura: Hylidae). *Occasional Papers. Museum of Natural History, Kansas* 87, 1–40.

Cantino, P. D. (1982). Affinities of the Lamiales: a cladistic analysis. *Systematic Botany* 7, 237–248.

Carlquist, S. (1976a). Wood anatomy of Roridulaceae: ecological and phylogenetic implications. *American Journal of Botany* 63, 1003–1008.

Carlquist, S. (1976b). Wood anatomy of Byblidaceae. *Botanical Gazette* 137, 35–8.

Carlton, C. E. (1990). Biogeographic affinities of pselaphid Beetles of the eastern United States. *Florida Entomologist* 73, 570–579.

Carlton, C. E., and Cox, R. T. (1990). A new species of *Arianops* from Central Arkansas and biogeographic implications of the interior highlands *Arianops* species (Coleoptera: Pselaphidae). *The Coleopterists' Bulletin* 44, 365–371.

Carolin, R. (1987). A review of the family Portulacaceae. Australian *Journal of Botany* 35, 383–412.

Carson, H. L. (1992). The Galapagos that were. *Nature* 355, 202–203.

Carson, H. L., and Clague, D. A. (1995). Geology and biogeography of Hawaii. In *Hawaiian Biogeography: evolution on a hot-spot archipelago* (eds. W. L. Wagner and V. A. Funk), pp. 14–29. Smithsonian Institution Press, Washington, DC.

Case, J. A. (1989). Antarctica: the effects of high latitude heterochroneity on the origin of the Australian marsupials. In *Origins and evolution of the Antarctic biota* (ed. J. A. Crane), pp. 217–226. Geological Society Special Publication 47.

Cernosvitov, L. (1936). Notes sur la distribution mondiale de quelques Oligochètes. *Vestnik Spolecnosti Zoologicke* 3, 16–19.

Chamberlain, V. E., and Lambert, R. St. J., (1985). Cordilleria: a newly defined Canadian microcontinent. *Nature* 314, 707–713.

Chesser, R. T. and Zink, R. M. (1994). Modes of speciation in birds: a test of Lynch's method. *Evolution* 48, 490–497.

Chiappe, L. M. (1995). The first 85 million years of avian evolution. *Nature 378, 349–355.*

Chiba, H. (1988). A lepidopterist's view of panbiogeography. *Rivista di Biologia—Biology Forum* 81, 553–568.

Chiba, H. (1989). Systematics and Panbiogeography. In *Dynamic structures in biology* (eds. B. Goodwin, A. Sibitani, and G. Webster), pp. 211–218. Edinburgh University Press, Edinburgh.

Chin, N. K. M., Brown, M. T., and Heads, M. J. (1991). The biogeography of Lessoniaceae, with special reference to *Macrocystis* C. Agardh (Phaeophyta: Laminariales). *Hydrobiologia* 215, 1–11.

Christiansen, K., and Culver, D. (1985). Biogeography and the distribution of cave Collembola. *Journal of Biogeography* 14, 459–477.

Christie, D. M., Duncan, R. A., McBirney, A. R., Richards, M. A., White, W. M., Harpp, K. S., and Fox, C. G. (1992). Drowned islands downstream from the Galapagos hotspot imply extended speciation times. *Nature* 355, 246–248.

Clegg, M. T. (1989). Dating the monocot-dicot divergence. *Trends in Ecology and Evolution* 15, 1–2.

Climo, F. M. (1990). The panbiogeography of New Zealand as illuminated by the genus *Fectola* Iredale 1915 and subfamily Rotadiscinae Pilsbury, 1927 (Mollusca: Pulmonata: Punctoidea: Charopidae). *New Zealand Journal of Zoology* 16, 587–649.

Cockburn, A. (1989). Trees, cows and cocaine: an interview with Susana Hecht. *New Left Review* 173:34–45.

Colacino, C. (1997). Léon Croizat's biogeography and macroevolution, or . . . "Out of nothing, nothing comes." *The Philippine Scientist* 34:73–88.

Connor, E. F. (1988). Fossils, phenetics and phylogenetics: inferring the historical dynamics of biogeographic distributions. In *Zoogeography of Caribbean Insects* (ed. J. K. Leibherr), pp. 254–269. Cornell University Press, Ithaca, New York.

Conroy, M. J., and Noon, B. R. (1996). Mapping of species richness for conservation of biological diversity: conceptual and methodological issues. *Ecological Applications* 6, 763–773.

Coope, G. R. (1970). Interpretations of Quaternary insect fossils. *Annual Review of Entomology* 15, 97–120.

Coope, G. R. (1978). Constancy of insect species versus inconstancy of Quaternary environments. In *Diversity of insect faunas* (eds. L. A. Mound and N. Waloff), pp. 176–187. Blackwell Scientific Publications, Oxford.

Coope, G. R. (1994). The response of insect faunas to glacial-interglacial climatic fluctuations. *Philosophical Transactions of the Royal Society of London B* 344, 19–26.

Corner, E. J. H. (1963). Ficus in the Pacific Ocean. In *Pacific basin biogeography* (ed. J. L. Gressitt), pp. 233–245. Bishop Museum Press, Honolulu.

Corner, E. J. H. (1985). Essays on *Ficus*. *Allertonia* 4, 125–166.

Cornet, B. (1986). The leaf venation and reproductive structures of a late Triassic angiosperm *Sanmiguelia lewisii*. *Evolutionary Theory* 7, 231–309.
Cornet, B. (1989). The reproductive morphology and biology of *Sanmiguelia lewisii* and its bearing on angiosperm evolution in the late Triassic. *Evolutionary Trends in Plants* 3, 25–51.
Cornet, B., and Habib, D. (1992). Angiosperm-like pollen from the ammonite-dated Oxfordian Upper Jurassic. *Review of Palaeobotany and Palynology* 71, 269–294.
Cowie, R. H. (1995). Variation in species-diversity and shell shape in Hawaiian land snails. In-situ speciation and ecological relationships. *Evolution* 49, 1191–1202.
Cox, C. B. (1990). New geological theories and old biogeographical problems. *Journal of Biogeography* 17, 117–130.
Cox, C. B., and Moore, P. D. (1993). *Biogeography. An ecological and evolutionary approach*. Blackwell Scientific Publications, Oxford.
Cracraft, J. (1975). Historical biogeography and earth history: perspectives for a future synthesis. *Annals of the Missouri Botanical Garden* 62, 227–250.
Cracraft, J. (1980). The biogeographic patterns of terrestrial SW Pacific vertebrates in the Southern Ocean. *Palaeogeography, Palaeoclimatology, Palaeoecology* 31, 353–369.
Cracraft, J. (1982). Phylogenetic relationships and trans-antarctic biogeography of some gruiform birds. *Géobios Mémoire Spécial* 6, 393–402.
Cracraft, J. (1985). Historical biogeography and patterns of differentiation within the South American avifauna: areas of endemism. *Ornithological Monographs* 36, 47–84.
Cracraft, J. (1986). Origin and evolution of continental biotas: speciation and historical congruence within the Australian avifauna. *Evolution* 40, 977–996.
Cracraft, J. (1991). Patterns of diversification within continental biotas: hierarchical congruence among the areas of endemism of Australian vertebrates. *Australian Systematic Botany* 4, 211–227.
Craig, A. J. F. K. (1985). The distribution of the pied starling, and southern African biogeography. *Ostrich* 56, 123–131.
Crame, J. A. (1992). Evolutionary history of the polar regions. *Historical Geology* 6, 37–60.
Crame, J. A. (1993). Bipolar molluscs and their evolutionary implications. *Journal of Biogeography* 20, 145–161.
Crane, P. R. (1985). Phylogenetic analysis of seed plants and the origin of angiosperms. *Annales of the Missouri Botanical Garden* 72, 716–793.
Cranston, P. S., and Edwards, D. H. D. (1992). A systematic reappraisal of the Australian Aphroteniinae (Diptera: Chironomidae) with dating from vicariance biogeography. *Systematic Entomology* 17, 41–54.
Cranston, P. S., Edwards, D. H. D., and Colless, D. H. (1987). *Archaeochlus* Brundin: a midge out of time (Diptera: Chironomidae). *Systematic Entomology* 12, 313–334.
Cranston, P. S., and Naumann, D. (1991). Biogeography. In *The insects of Australia* (ed. I. D. Naumann), pp. 180–197. Melbourne University Press, Melbourne.

Craw, R. C. (1978a). Two biogeographical frameworks: implications for the biogeography of New Zealand. A review. *Tuatara* 23, 81–114.
Craw, R. C. (1978b). Revision of *Argyrophenga* (Lepidoptera: Satyridae). *New Zealand Journal of Zoology* 5, 751–768.
Craw, R. C. (1979). Generalized tracks and dispersal in biogeography: a response to R. M. McDowall. *Systematic Zoology* 28, 99–107.
Craw, R. C. (1982). Phylogenetics, areas, geology and the biogeography of Croizat: a radical view. *Systematic Zoology* 31, 304–316.
Craw, R. C. (1983). Panbiogeography and vicariance biogeography: are they truly different? *Systematic Zoology* 32, 431–438.
Craw, R. C. (1984). Leon Croizat's biogeographic work: a personal appreciation. *Tuatara* 27, 8–13.
Craw, R. C. (1985). Classic problems of southern hemisphere biogeography re-examined. *Zeitschrift für Zoologische Systematik und Evolutionsforschung* 23, 1–10.
Craw, R. C. (1988). Panbiogeography: method and synthesis in biogeography. In *Analytical biogeography* (ed. A. A. Myers and P. S. Giller), pp. 437–481. Chapman and Hall, London.
Craw, R. C. (1989). Continuing the synthesis between panbiogeography, phylogenetic systematics and geology as illustrated by empirical studies on the biogeography of New Zealand and the Chatham Islands. *Systematic Zoology* 37, 291–310.
Craw, R. C. (1990). New Zealand biogeography: a panbiogeographic approach. *New Zealand Journal of Zoology* 16, 527–547.
Craw, R. C. (1993). Anthropophagy of the other. *Art and Asia Pacific* 1, 10–15.
Craw, R. C., and Hubbard, G. (1993). Cross pollination: hyphenated identities and hybrid realities (or ALTER/NATIVE to what?). *Midwest* 3, 32–33.
Craw, R. C., and Page, R. (1988). Panbiogeography: method and metaphor in the new biogeography. In *Evolutionary processes and metaphors* (eds. M.-W. Ho and S. Fox), pp. 163–189. John Wiley and Sons, Chichester, UK.
Craw, R. C., and Weston, P. (1984). Panbiogeography: a progressive research programme? *Systematic Zoology* 33, 1–13.
Crisci, J. V. (1980). Evolution in the subtribe Nassauviinae (Compositae, Mutisieae): a phylogenetic reconstruction. *Taxon* 29, 213–224.
Crisci, J. V., and Stuessy, T. F. (1980). Determining primitive character states for phylogenetic reconstruction. *Systematic Botany* 5, 112–236.
Croizat, L. (1943). Notes on the Cactaceae: the typification of *Echinocactus*. *Lilloa* 9, 179–198.
Croizat, L. (1952). *Manual of phytogeography*. Junk, The Hague.
Croizat, L. (1958). *Panbiogeography*. [Published by the author], Caracas.
Croizat, L. (1961). *Principia Botanica*. [Published by the author], Caracas.
Croizat, L. (1964). *Space, time, form: the biological synthesis*. [Published by the author], Caracas.
Croizat, L. (1968a). The biogeography of the tropical lands and islands of Suez-Madagascar: with particular reference to the dispersal and form-making of *Ficus* L., and different other vegetal and animal groups. *Atti*

Istituto Botanico Universita Laboratorio Crittogamico. Pavia, Serie 6, 4, 1–400.
Croizat, L. (1968b). Introduction raisonée à la biogéographie de l'Afrique. *Memórias da Sociedade Broteriana* 20, 1–451.
Croizat, L. (1971). De la "pseudovicariance" et de la "disjonction illusoire". *Anuario da Sociedade Broteriana* 37, 113–140.
Croizat, L. (1976). *Biogeografía analítica y sintética ("Panbiogeografía") de las Américas*. Biblioteca Acad. Cienc. Fis. Mat. Nat., Caracas.
Croizat, L. (1993). On the structural and developmental history of the capsule of *Mesembryanthemum s.l. Kirkia* 14, 145–169.
Croizat, L. (1994). Observations on the biogeography of the genus *Goliathus* (*Insecta*: *Coleoptera*). *Kirkia 15 (2)*: 141–155.
Crovello. T. J. (1981). Quantitative biogeography: an overview. *Taxon* 30, 563–575.
Crow, J. F. (1987). Some basic principles of organismic and molecular evolution. In *Evolution and vertebrate immunity* (eds. G. K. Schulze and D. H. Schulze), pp. 1–14. University of Texas Press, Austin.
Crozier, R. H. (1992) Genetic diversity and the agony of choice. *Biological Conservation* 61, 11–15.
Cumming, J. M. (1989). Classification and evolution of the eumenine wasp genus *Symmorphus* Wesmael (Hymenoptera: Vespidae). *Memoirs of the Entomological Society of Canada* 148.
Cunningham, C. W., Buss, L. W., and Anderson, C. (1991). Molecular and geologic evidence of shared history between hermit crabs and the symbiotic genus *Hydractinia*. *Evolution* 45, 1301–1315.
Cunningham, C. W., and Collins, T. M. (1994). Developing model systems for molecular biogeography: vicariance and interchange in marine invertebrates. In *Molecular ecology and evolution: approaches and applications* (eds. B. Schierwater, B. Streit, G. P. Wagner, and R. deSalle), pp. 405–433. Birkhauser Verlag, Basel.
Darlington, P. J. (1957). *Zoogeography: the geographic distribution of animals*. Wiley, New York.
Darwin, C. (1859 [1911]). *On the origin of species by means of natural selection or the preservation of favoured races in the struggle for life*. John Murray, London.
Darwin, C. (1860). *On the origin of species by means of natural selection*. 5th Thousand. John Murray, London.
Darwin, C. (1888). *The descent of man and selection in relation to sex*. 2nd edition. John Murray, London.
Darwin, F., and Seaward, A. C. (1903). *More letters of Charles Darwin*. John Murray, London.
Davis, A. L. V. (1993). Biogeographical groups in a southern African, winter rainfall, dung beetle assemblage (Coleoptera: Scarabaeidae)—consequences of climatic history and habitat fragmentation. *African Journal of Ecology* 31, 306–327.
Davis, D. R. (1994). Neotropical Tineidae V: the Tineidae of Cocos Island, Costa Rica (Lepidoptera: Tineoidea). *Proceedings of the Entomological Society of Washington*. 96, 735–748.
Davis, P. H., and Heywood, V. H. (1963). *Principles of angiosperm taxonomy*. Oliver and Boyd, Edinburgh.

de Candolle, A. -P. (1855). *Géographie botanique raisonnée*, vols. 1 and 2. J. Kressman, Geneva.

de Candolle, A. -P. (1886 [1959]). *Origin of cultivated plants* [reprint of 2nd ed.]. Hafner Publishing Co., New York.

De Lattin, G. (1967). *Grundriss der Zoogeographie*. Gustav Fischer Verlag, Stuttgart.

de Jong, R. (1980). Some tools for evolutionary and phylogenetic studies. *Zeitschrift für Zoologische Systematik und Evolutionsforschung* 18, 1–23.

de Queiroz, K. (1987). Phylogenetic systematics of iguanine lizards. *University of California Publications in Zoology* 118, 1–203.

de Queiroz, K., Kim, J., and Donoghue, M. J. (1995). Separate versus combined analysis of phylogenetic evidence. *Annual Review of Ecology and Systematics* 26, 657–681.

De Vries, E. J. (1985). The biogeography of the genus *Dugesia* (Turbellaria, Tricladida, Paludicola) in the Mediterranean region. *Journal of Biogeography* 12, 509–518.

De Weerdt, W. H. (1989). Phylogeny and vicariance biogeography of North Atlantic Chalinidae (Haplosclerida, Demospongieae). *Beaufortia* 39, 55–88.

De Weerdt, W. H. (1990). Discontinuous distribution of the tropical West Atlantic hydrocoral *Millepora squarrosa*. *Beaufortia* 41, 195–203.

Deevey, E. S. (1949). Biogeography of the Pleistocene I. Europe and North America. *Bulletin of the Geological Society of America* 60, 1315–1416.

Del Hoyo, and Elliott, A., and Sargata, J. (1994). *Handbook of the birds of the world*, vol 2. Lynx Edicions, Barcelona.

Dennis, R. L. H. (1995). *Butterflies and climate change*. Manchester University Press, Manchester, UK.

Desmond, A., and Moore, J. (1992). *Darwin*. Penguin, London.

Di Marco, G. (1994). Les terrains accrétés du sud du Costa Rica: Évolution tectonostratigraphique de la marge occidentale de la plaque Caraibe. *Memoire de Géologie (Lausanne)* 20, 1–184.

Diels, L. (1924). Die methoden der phytogeographie und der systematik der pflanzen. *Abderhalden's Handbuch der biologie Arbeitsmethoden* 11, 1:67–190.

Dobson, J. E. (1992). Spatial logic in paleogeography and the explanation of continental drift. *Annals of the Association of American Geographers* 82, 187–206.

Doyle, J. A., and Donoghue, M. J. (1986). Seed plant phylogeny and the origin of angiosperms: an experimental cladistic approach. Botanical Review 52, 321–431.

Doyle, J. A., and Donoghue, M. J. (1987). The origin of angiosperms. In *The origin of angiosperms and their biological consequences* (eds. E. M. Friis, W. G. Chaloner, and P. R. Crane), pp. 17–49. Cambridge University Press, Cambridge.

Dover, G. A. (1982). Molecular drive: a cohesive mode of species evolution. *Nature* 299, 111–116.

Dover, G. A. (1986). The spread and success of non-Darwinian novelties. In *Evolutionary processes and theory* (eds. S. Karlin and E. Nevo), pp. 199–237. Academic Press, Orlando, Florida.

Dover, G. A. (1992). Observing development through evolutionary eyes: a practical approach. *Bioessays* 14, 281–287.

Dover, G. A. (1993). Evolution of genetic redundancy for advanced players. *Current Opinion in Genetics and Development* 3, 902–910.

Dover, G. A., and Strachan, T. (1987). Molecular drive in the evolution of immune superfamily genes: the initiation and spread of novelty. In *Evolution and vertebrate community* (eds. G. K. and D. H. Schulze), pp. 15–33. University of Texas Press, Austin.

Dover, G. A., Linares, A. R., Bowen, T., and Hancock, J. M. (1993). Detection and quantification of concerted evolution and molecular drive. *Methods in Enzymology* 224, 631–646.

Duellman, W. E. (1986). Plate tectonics, phylogenetic systematics, and vicariance biogeography of Anura: methodology for unresolved problems. In *Studies in Herpetology* (ed. Z. Rocek), pp. 59–62. Charles University, Prague.

Duellman, W. E., and Campbell, J. A. (1992). Hylid frogs of the genus *Plectrohyla*: systematics and phylogenetic relationships. *Miscellaneous Publications Museum of Zoology University of Michigan* 181.

Duellman, W. E., and Veloso, M. A. (1977). Phylogeny of *Pleurodema* (Anura: Leptodactylidae): a biogeographic model. *Occasional Papers. Museum of Natural History, Kansas* 64, 1–46.

Dugdale, J. S. (1994). Hepialidae (Insecta: Lepidoptera). *Fauna of New Zealand* 30, 1–163.

Duncan, R. A., and Hargraves, R. B. (1984). Plate tectonic evolution of the Caribbean region in the mantle reference frame. *Geological Society of America Memoir* 162, 81–93.

Durham, J. W. (1985). Movement of the Caribbean plate and its importance for biogeography in the Caribbean. *Geology* 13, 123–125.

Easteal, S., Collet, C., and Betty, D., eds. (1995). *The mammalian molecular clock.* Springer-Verlag, New York.

Ehrlich, P. R. (1992) Population biology of checkerspot butterflies and the preservation of global biodiversity. *Oikos* 63, 6–12.

Elder, J. F. Jr., and Turner, B. J. (1994). Concerted evolution at the population level: pupfish *Hind*III satellite DNA sequences. *Proceedings of the National Academy of Sciences USA* 91, 994–998.

Elder, J. F. Jr., and Turner, B. J. (1995). Concerted evolution of repetitive DNA sequences. *The Quarterly Review of Biology* 70, 297–320.

Eldredge, N., ed. (1992). *Systematics, ecology, and the biodiversity crisis.* Columbia University Press, New York.

Engler, A. (1882). *Versuch einer Entwicklungsgeschichte der Pflanzenwelt, insbesondere der florengebiete seit der Tertiärperiode.* 2 Teile. Englemann, Leipzig.

Ereshefsky, M., ed. (1992). *The units of evolution: essays on the nature of species.* MIT Press, Cambridge, Massachusetts.

Eskov, K. Y., and Golovatch, S. I. (1986). On the origin of trans-Pacific disjunctions. *Zoologische Jahrbücher Systematik* 113, 265–285.

Espinosa, D., and Llorente, J. (1993). *Fundamentos de Biogeografías Filogenéticas.* Universidad Nacional Autónoma de México and CONABIO, México, D. F.

Estabrook, G. F., and Gates, B. (1984). Character analysis in the *Bannisteriopsis campestris* complex (Malpighiaceae), using spatial auto-correlation. *Taxon* 33, 13–25.

Exell, A. W., and Stace, C. A. (1972). Patterns of distribution in the Combretaceae. In *Taxonomy, Phytogeography, and Evolution* (ed. D. H. Valentine), pp. 307–323. Academic Press, London.

Faith, D. P. (1994). Phylogenetic pattern and the quantification of organismal biodiversity. *Philosophical Transactions of the Royal Society of London B* 345, 45–58.

Farris, J. S. (1970). Methods for computing Wagner trees. *Systematic Zoology* 19, 83–92.

Feininger, T. (1987). Allochthonous terranes in the Andes of Ecuador and northwestern Peru. *Canadian Journal of Earth Sciences* 24, 266–278.

Feldmann, R. M., and Wilson, M. T. (1988). Eocene decapod crustaceans from Antarctica. *Geological Society of America Memoir* 169, 465–488.

Felsenstein, J. (1993). *PHYLIP 3.5*. University of Washington, Seattle.

Ferris, V. R. (1979). Cladistic approaches in the study of soil and plant parasitic nematodes. *American Zoologist* 19, 1195–1215.

Ferris, V. (1980). A science in search of a paradigm? Review of the symposium "Vicariance biogeography: a critique". *Systematic Zoology* 29, 67–76.

Field, W. D. (1971). Butterflies of the genus *Vanessa* and of the resurrected genera *Bassaris* and *Cynthia* (Lepidoptera: Nymphalidae). *Smithsonian Contributions to Zoology* 84, 1–75.

Fisher, D. C. (1994). Stratocladistics: morphological and temporal patterns and their relation to phylogenetic processes In *Interpreting the hierarchy of nature* (eds. L. Grande and O. R. Rieppel), pp. 133–172. Academic Press, San Diego.

Flint, O. S. (1978). Probable origins of the West Indian Trichoptera and Odonata faunas. In *Proceedings of the 2nd International Symposium on Trichoptera*, pp. 215–223. Junk, The Hague.

Fjeldsa, J. (1994). Geographical patterns for relict and young species of birds in Africa and South America and implications for conservation priorities. *Biodiversity and Conservation* 3, 207–226.

Forman, L. L. (1966). *On the evolution of cupules in the Fagaceae*. Kew Bulletin 18, 383–419.

Fortes, M. D. (1988). Indo-West Pacific affinities of Philippine seagrasses. *Botanica Marina* 31, 237–242.

Fortíno, A. D., and J. J. Morrone (1997). Signos graficos para la representacion de analisis panbiogeograficos. *Biogeographica 73*: 49–56.

Fosberg, F. R. (1976). Geography, ecology and biogeography. *Annals of the Association of American Geography* 66, 117–128.

Fosberg, F., and Sachet, M.-H. (1972). *Thespesia populnea* and *T. populneoides* (Malvacae). *Smithsonian Contributions to Botany* 7, 1–13.

Francke, O. F., and Soleglad, M. E. (1981). The family Luridae Thorell (Arachnida, Scorpiones). *Journal of Arachnology* 9, 233–258.

Fry, C. H., Keith, S., and Urban, E. K. (1988). *The birds of Africa*. vol. 3. Academic Press, London.

Fryxell, P. A. (1979).*The natural history of the cotton tribe (Malvaceae, tribe Gossypieae)*. Texas A&M University Press, College Station.

Furon, R. (1963). *Geology of Africa*. Oliver and Boyd, London.
Galloway, D. J. (1988). Plate tectonics and the distribution of cool temperate southern hemisphere macrolichens. *Botanical Journal of the Linnean Society* 96, 45–55.
Garrison, R. W. (1992). Using ordination methods with geographic information: species resolution in a partially sympatric complex of neotropical *Tramea* dragonflies (Odonata: Libellulidae). In *Ordination in the study of morphology, evolution and systematics of insects: applications and quantitative genetic ratios* (eds. J. T. Sorensen and R. Foottit), pp. 223–240. Elsevier Science Publishers, Amsterdam.
Gascoigne, A. (1994). The biogeography of land snails in the islands of the Gulf of Guinea. *Biodiversity and Conservation* 3, 794–807.
Gastil, R. G., and Jensky, W. (1973). Evidence for strike-slip displacement beneath the trans-Mexican volcanic belt. In *Proceedings of the conference on tectonic problems of the San Andreas fault system* (eds. R. L. Kovach and A. Nur), pp. 171–180. School of Earth Sciences, Stanford University, Stanford, California.
George, W. (1993). Appendix: the history of the problem. In *The Africa-South America Connection* (eds. W. George and R. Larocat), pp. 151–161. Clarendon Press, Oxford.
Ghiold, J., and Hoffman, A. (1984). Clypeasteroid echinoids and historical biogeography. *Neues Jahrbuch für palaeontologie Monatschefte* 9, 529–538.
Ghiselin, M. T. (1974). A radical solution to the species problem. *Systematic Zoology* 23, 536–544.
Ghiselin, M. T. (1980). The failure of morphology to assimilate Darwinism. In *The evolutionary synthesis* (eds. E. Mayr and W. B. Provine), pp. 180–193. Harvard University Press, Cambridge, Massachusetts.
Gibbs, G. W. (1980). *New Zealand butterflies: identification and natural history*. Collins, Auckland.
Gibbs, G. W. (1983). Evolution of Micropterigidae (Lepidoptera) in the SW Pacific. *Geojournal* 7, 505–510.
Gibbs, G. W. (1990). Local or global? Biogeography of some primitive Lepidoptera in New Zealand. *New Zealand Journal of Zoology* 16, 689–698.
Gillett, J. (1966). Notes on Leguminosae (Phaseoleae). *Kew Bulletin* 20, 103–112.
Gillett, J. B. (1991). Burseraceae. In *Flora of tropical East Africa* (ed. R. M. Polhill), pp. 1–94. A.A. Balkema, Rotterdam.
Glaubrecht, M. (1992). Lange evolution auf Galápagos. *Naturwissenschaftliche Rundschau* 45, 404–405.
Good, R. (1964). *The geography of flowering plants*. Longmans, London.
Gotelli, N. J. (1995). *A primer of ecology*. Sinauer Associates, Sunderland, Massachusetts.
Grande, L. (1985). The use of paleontology in systematics and biogeography, and a time control refinement for historical biogeography. *Paleobiology* 11, 234–243.
Grande, L. (1989). The Eocene Green River lake system, fossil lake, and the history of the North American fish fauna. In *Mesozoic/Cenozoic vertebrate paleontology: classic localities, contemporary approaches* (ed. J. J. Flynn), pp. 18–28. American Geophysical Union, Washington, DC.

Grande, L. (1990). Vicariance biogeography. In *Palaeobiology: a synthesis* (eds. D. E. G. Briggs and P. R. Crowther), pp. 448–451. Blackwell Scientific Publications, Oxford.

Grant, P. R., and B. R. Grant. (1992). Hybridization of bird species. *Science* 256, 193–197.

Gray, R. D. (1988). Metaphors and methods: behavioural ecology, panbiogeography and the evolving synthesis. In *Evolutionary Processes and Metaphors* (eds. M.-W. Ho and S. Fox), pp. 209–242. John Wiley and Sons, Chichester, UK.

Gray, R. D. (1990). Opposition in panbiogeography: can the conflicts between selection, constraint, ecology, and history be resolved? *New Zealand Journal of Zoology* 16, 787–806.

Gray, R. D. (1992). Death of the gene: developmental systems strike back. In *Trees of life: essays in philosophy of biology* (ed. P. Griffiths), pp. 165–209. Kluwer Academic Publishers, Dordrect.

Grehan, J. R. (1984). Evolution by law: Croizat's 'orthogeny' and Darwin's 'laws of growth'. *Tuatara* 27, 14–19.

Grehan, J. R. (1987). Evolution of arboreal tunneling by larvae of *Aenetus* (Lepidoptera: Hepialidae). *New Zealand Journal of Zoology* 14, 441–462.

Grehan, J. R. (1990a). Panbiogeography and conservation science in New Zealand. *New Zealand Journal of Zoology* 16, 731–748.

Grehan, J. R, (1990b). Invertebrate survey of Somes Island (Matiu) and Mokopuna Island, Wellington Harbour, New Zealand. *New Zealand Entomologist* 13, 62–75.

Grehan, J. R. (1991a). A panbiogeographic perspective for pre-Cretaceous angiosperm-Lepidoptera coevolution. *Australian Systematic Botany* 4, 91–110.

Grehan, J. R. (1991b). Panbiogeography 1981–91: development of an earth/life synthesis. *Progress in Physical Geography* 15, 331–363.

Grehan, J. R. (1992). Biogeography and conservation in the real world. *Global Ecology and Biogeography Letters* 2, 96–97.

Grehan, J. R. (1993). Conservation biogeography and the biodiversity crisis: a global problem in space/time. *Biodiversity Letters* 1, 134–140.

Grehan, J. R. (1994). The beginning and end of dispersal: the representation of 'panbiogeography'. *Journal of Biogeography* 21, 451–462.

Grehan, J. R. (1995). Natural biogeographic patterns of biodiversity: the research imperative. In *Proceedings of the second international conference on science and the management of protected areas. Nova Scotia 1994* (eds. T. B. Herman, S. Bondrup-Nielsen, J. H. M. Willison, and N. W. P. Munro), pp. 35–44. Science and Management of Protected Areas Association, Canada.

Grehan, J. R., and Ainsworth, R. (1985). Orthogenesis and evolution. *Systematic Zoology* 34, 174–192.

Grehan, J. R., and Climo, F. (1992). Conservation biogeography of North-West Nelson, New Zealand. Unpublished report, New Zealand Conservation Authority, Wellington.

Grene, M. (1990). Is evolution at the cross-roads? *Evolutionary Biology* 24, 51–58.

Griffiths, P. E., ed. (1992). *Trees of life: essays in the philosophy of biology.* Kluwer Academic Publishers, Dordrecht.

Griffiths, P. E., and Gray, R. D. (1994). Developmental systems and evolutionary systems. *Journal of Philosophy 91*: 227–304.

Grubb, P. (1978). Patterns of speciation in African mammals. *Bulletin of the Carnegie Museum of Natural History* 6, 152–67.

Guiry, M. D., and D. J. Gabary. (1990). A preliminary phylogenetic analysis of the Phyllophoraceae, Gigantinaceae and Petrocelidaceae (Rhodophyta) in the North Atlantic and North Pacific. In *Evolutionary biogeography of the marine algae of the North Atlantic* (eds. D. J. Garbury and G. Robin South), pp. 265–290. Springer-Verlag, Berlin.

Haddock, M. H. (1938). *Disrupted strata: faulting and its allied problems from the standpoint of the mine surveyor and stratigraphist,* 2nd ed. The Technical Press Ltd., London.

Haffer, J. (1969). Speciation in Amazonian forest birds. *Science* 165, 131–137.

Hale, M. (1975). A monograph of the lichen genus *Relicina* (Parmeliaceae). *Smithsonian Contributions in Botany* 26, 1–32.

Hall, B. P., and Moreau, R. E. (1970). *An atlas of speciation in African passerine birds.* Trustees of the British Museum (Natural History), London.

Hallam, A. (1986). Evidence of displaced terranes from Permian to Jurassic faunas around the Pacific margins. *Journal of the Geological Society London* 143, 209–216.

Hallam, A. (1994). *An outline of Phanerozoic Biogeography*. Oxford University Press, Oxford.

Hamilton, S. W., and Morse, J. C. (1990). Southeastern caddisfly fauna: origins and affinities. *Florida Entomologist* 73, 587–600.

Hammond, P. C. (1991). Patterns of geographic variation and evolution in polytypic butterflies. *Journal of Research on the Lepidoptera* 29, 54–76.

Hammond, P. M. (1994). Practical approaches to the estimation of the extent of biodiversity in speciose groups. *Philosophical Transactions of the Royal Society of London B* 345, 119–136.

Hanafusa, H. (1992). Three new nymphalid butterflies from Indonesia and Philippines. *Futao* 10, 1–2, 7–8, 15–16.

Harms, H., (1940). Meliaceae. *Die natürlichen Pflanzenfamilien 2 Aufl.* 19b, 1–172.

Harold, A. S., and Telford, M. (1990). Systematics, phylogeny and biogeography of the genus *Mellita* (Echinoidea: Clypeasteroidea). *Journal of Natural History* 24, 987–1026.

Harper, J. L., and Hawksworth, D. L. (1994). Biodiversity: measurement and estimation. Preface. *Philosophical Transactions of the Royal Society of London B* 345, 5–12.

Harris, A. C. (1983). An Eocene larval insect fossil (Diptera: Bibionidae) from North Otago, New Zealand, with an appendix by Aitchison, J. C., Campbell, H. J., Campbell, S. D., and Raine, J. I. *Journal of the Royal Society of New Zealand* 13, 93–105.

Harrison, A. D., and Rankin, J. J. (1976). Hydrobiological studies of eastern Lesser Antillean Islands II: St. Vincent: freshwater fauna—its distribution, tropical river zonation and biogeography. *Archiv für Hydrobiologie Supplementband* 50, 275–311.

Hayami, I. (1989). Outlook on the post-Paleozoic historical biogeography of pectinids in the Western Pacific region. In *Current aspects of biogeography in west Pacific and east Asia regions* (eds. H. Ohba, I. Hayami, and K. Mochizuki), pp. 3–25. The University Museum, University of Tokyo, Tokyo.

Haydon, D. T., Crother, B. I., and Pianka, E. R. (1994a). New directions in biogeography. *Trends in Ecology and Evolution* 10, 403–406.

Haydon, D. T., Radtkey, R. R., and Pianka, E. R. (1994b). Experimental biogeography: interactions between stochastic, historical, and ecological processes in a model archipelago. In *Species diversity in ecological communities: historical and geographical perspectives* (eds. R. E. Ricklefs and D. Schluter), pp. 267–280. University of Chicago Press, Chicago.

Heads, M. (1983). Book Review. *Journal of Biogeography 10*: 543–548.

Heads, M. (1984). *Principia Botanica*: Croizat's contribution to botany. *Tuatara 27*: 26–48.

Heads, M. J. (1985). On the nature of ancestors. *Systematic Zoology* 34, 205–215.

Heads, M. J. (1990). Integrating earth and life sciences in New Zealand natural history: the parallel arcs model. *New Zealand Journal of Zoology* 16, 549–585.

Heads, M. (1991). A revision of the genera *Kelleria* and *Drapetes* (Thymeliaceae). *Australian Systematic Botany* 3, 595–652.

Heads, M. J. (1993). Biogeography and biodiversity in *Hebe*, a South Pacific genus of Scrophulariaceae. *Candollea* 48, 19–60.

Heads, M. J. (1994a). A biogeographic review of *Parahebe* (Scrophulariaceae). *Botanical Journal of the Linnean Society of London* 115, 65–89.

Heads, M. J. (1994b). Biogeographic studies in New Zealand Scrophulariaceae: tribes Rhinantheae, Calceolarieae and Gratioleae. *Candollea* 49, 55–80.

Heads, M. J. (1994c). Biogeography and evolution in the *Hebe* complex (Scrophulariaceae): *Leonohebe* and *Chionohebe*. *Candollea* 49, 81–119.

Heads, M. J. (1994d). Biogeography and biodiversity in New Zealand *Pimelea* (Thymelaeaceae). *Candollea* 49, 37–53.

Heads, M. J. (1997). Regional patterns of biodiversity in New Zealand: one degree grid analysis of plant and animal distributions. *Journal of the Royal Society of New Zealand* 27: 337–354.

Heads, M. J. (1998). Biogeographic disjunction along the Alpine fault, New Zealand. *Biological Journal of the Linnean Society 63*: 161–176.

Hedges, S. B. (1986). An electrophoretic analysis of Holarctic hylid frog evolution. *Systematic Zoology* 35, 1–21.

Hedges, S. B. et al. (1996). Continental breakup and the ordinal diversification of birds and mammals. *Nature* 381, 226–229.

Henderson, I. (1985). *Systematic studies of New Zealand Trichoptera and critical analysis of systematic methods* (Ph.D. thesis). Victoria University of Wellington.

Henderson, I. (1990). Quantitative biogeography: an investigation into concepts and methods. *New Zealand Journal of Zoology* 16, 495–510.

Henderson, I. (1991). Biogeography without area? *Australian Systematic Botany* 4, 59–71.

Hendrickson, D. A. (1986). Congruence of bolitoglossine biogeography and phylogeny with geologic history: palaeotransport on displaced suspect terranes. *Cladistics* 2, 113–129.

Hennig, W. (1950). *Grundzüge einer theorie der phylogenetischen systematik.* Deutscher Zentraverlag, Berlin.

Hennig, W. (1966). *Phylogenetic systematics.* University of Illinois Press, Urbana.

Herenden, P. S., Les, D. H., and Dilcher, D. L. (1990). Fossil *Ceratophyllum* Ceratophyllaceae from the Tertiary of North America. American Journal of Botany 77, 7–16.

Heslop-Harrison, J. (1983). The reproductive versatility of flowering plants: an overview. In *Strategies of plant reproduction* (ed. W. Meudt), pp. 3–18. Allanheld Osmun, London.

Hesse, R. (1913). Die ökologischen grandlagen der Tierverbreitung. *Geographische Zeitzschrift (Leipzig)* 19, 241–259, 335–345, 445–460, 498–513.

Hesse, R. (1924). *Tiergeographie auf oekologischer grundlage.* Fischer, Jena.

Hese, R., Allee, W. C., and Schmidt, K. P. (1937). *Ecological animal geography.* Wiley, New York.

Heyer, W. R. (1975). A preliminary analysis of the intergeneric relationships of the frog family Leptodactylidae. *Smithsonian Contributions to Zoology* 199.

Heyer, W. R., and Maxson, L. R. (1982). Neotropical frog biogeography: paradigms and problems. *American Zoologist* 22, 397–410.

Hibbett, D. S., Grimaldi, D., and Donoghue, M. J. (1995). Cretaceous mushrooms. *Nature* 377, 487.

Hill, R. S. (1994). The history of selected Australian taxa. In *History of the Australian vegetation* (ed. R. S. Hill), pp. 390–419. Cambridge University Press, Cambridge (England), New York.

Hillis, D. M. (1985). Evolutionary genetics of the andean lizard genus *Pholidobolus* (Sauria: Gymnopthalmidae): phylogeny, biogeography, and a comparison of tree construction techniques. *Systematic Zoology* 34, 109–126.

Ho, J. (1988). Cladistics of *Sunaristes*, a genus of harpacticoid copepods associated with hermit crabs. *Hydrobiologia* 167/168, 555–560.

Ho, M. W. (1988). On not holding nature still: evolution by process, not by consequence. In *Evolutionary Processes and Metaphors* (eds. M. W. Ho and S. Fox), pp. 117–144. John Wiley and Sons, Chichester, U. K.

Hoberg, E. P., and Adams, A. M. (1992). Phylogeny, historical biogeography, and ecology of *Anophryocephalus* spp. (Encestoidea: Tetrabothriidae) among pinnipeds of the Holarctic during the late Tertiary and Pleistocene. *Canadian Journal of Zoology* 70, 703–719.

Hoelzer, G. A., Hoelzer, M. A., and Melnick, D. J. (1992). The evolutionary history of the *sinica* group of macaque monkeys as revealed by mtDNA restriction site analysis. *Molecular Phylogeny and Evolution* 1, 215–222.

Holden, J. C., and Dietz, R. S. (1972). Galapagos Gore, NazCoPac triple junction and Carnegie/Cocos ridges. *Nature* 235, 266–269.

Holdhaus, K. (1954). *Die Spuren der Eiszeit in der Tierwelt Europas.* Wagner, Insbruck.

Holland, S. M. (1995). The stratigraphic distribution of fossils. *Paleobiology* 21, 92–109.

Holmes, A. (1944). *Principles of physical geology*. Thomas Nelson and Sons Ltd., London.
Holsinger, J. R. (1986). Zoogeographic patterns of North American subterranean amphipod crustaceans. In *Crustacean biogeography. Crustacean issues 3* (eds. R. H. Gore and K. L. Heck), pp. 85–106. Balkema, Rotterdam.
Holsinger, J. R. (1989). Preliminary zoogeographic analysis of five groups of crustaceans from anchialine caves in the West Indian region. *International Congress of Speleology* 10, 25–26.
Holsinger, J. R. (1991). What can vicariance biogeographic models tell us about the distributional history of subterranean amphipods? *Hydrobiologia* 223, 43–45.
Holsinger, J. R. (1994). Pattern and process in the biogeography of subterranean amphipods. *Hydrobiologia* 287, 131–145.
Holzenthal, R. W. (1986). Studies in neotropical Leptoceridae (Trichoptera), VI: Immature stages of *Hudsonema flaminii* (Navas) and the evolution and historical biogeography of Hudsonemini (Triplectidinae). *Proceedings of the Entomological Society of Washington* 88, 286–279.
Hooker, J. D. (1853). *Flora Novae-Zelandiae*. Lovell Reeve, London.
Höss, M., Dilling, A., Currant, A., and Pääbo, S. (1996). Molecular phylogeny of the extinct ground sloth *Mylodon darwinii*. *Proceedings of the National Academy of Science USA* 93, 181–185.
Houde, P. (1994). Evolution of the Heliornithidae: reciprocal illumination by morphology, biogeography, and DNA hybridization (Aves: Gruiformes). *Cladistics* 10, 1–19.
Hovenkamp, P. (1986). *A monograph of the fern genus Pyrrosia*. Leiden University Press, Leiden.
Howes, G. J. (1984). Phyletics and biogeography of the aspinine cyprinid fishes. *Bulletin of the British Museum of Natural History* (Zoology) 47, 283–303.
Howes, G. (1990). Biogeography of gadoid fishes. *Journal of Biogeography* 18, 595–622.
Howes, G. (1991). The syncranial osteology of the southern eel-cod family Muraenolepididae with comments on its phylogenetic relationships and on the biogeography of sub-Antarctic gadoid fishes. *Zoological Journal of the Linnean Society* 100, 73–100.
Hugot, J. P. (1982). Sur le genre *Wellcomia* (Oxyuridae, Nematoda) parasite de rongeurs archaiques. *Bulletin du Muséum National d'Histoire Naturelle (Paris), Serie* 4 (A), 25–48.
Hull, D. L. (1976). Are species really individuals? *Systematic Zoology* 25, 174–191.
Hull, D. L. (1988). *Science as a process: an evolutionary account of the social and conceptual development of science.* University of Chicago Press, Chicago.
Humphries, C. J. (1981). Biogeographical methods and the southern beeches (Fagaceae: *Nothofagus*). *Advances in cladistics* 1: 177–207.
Humphries, C. J. (1985). Review. *Journal of Natural History* 19, 1285–1286.
Humphries, C. J. (1990). The importance of Wallacea to biogeographical thinking. In *Insects and the rain forests of South East Asia (Wallacea)* (eds. W. J. Knight, and J. D. Holloway), pp. 7–18. Royal Entomological Society of London.

Humphries, C. J. (1992). Cladistic biogeography. In *Cladistics: a practical course in systematics* (ed. L. Forey), pp. 137–159. Clarendon Press, Oxford.

Humphries, C. J., Cox, J. M., and Nielsen, E. S. (1986). *Nothofagus* and its parasites: a cladistic approach to coevolution. In *Coevolution and systematics* (eds. A. R. Stone and D. L. Hawksworth), pp. 55–74. Clarendon Press, Oxford.

Humphries, C. J., and Parenti, L. (1986). *Cladistic biogeography.* Clarenden Press, Oxford.

Humphries, C. J., and Seberg, O. (1989). Graphs and generalized tracks: some comments on method. *Systematic Zoology* 38, 69–76.

Humphries, C. J., Williams, P. H., and Vane-Wright, R. I. (1995). Measuring biodiversity value for conservation. *Annual Review of Ecology and Systematics* 26, 93–111.

Hutchinson, J. (1946). *A botanist in Southern Africa*. Gawthorn: London.

Hutchinson, J. B., Silow, R. A., and Stephens, S. G. (1947). *The evolution of Gossypium and the differentiation of the cultivated cottons*. Oxford University Press, London.

Huxley, T. H. (1868). On the classification and distribution of the Alectoromorphidae and Heteromorphae. *Proceedings of the Zoological Society of London* 1868, 294–319.

Jablonski, D. (1987). Heritability at the species level: an analysis of geographic ranges of Cretaceous molluscs. *Science* 239, 360–363.

Jablonski, D., and Valentine, J. W. (1990). From regional to total geographic ranges: testing the relationship in Recent bivalves. *Paleobiology* 16, 126–142.

Jacobs, S. C., Larson, A., and Cheverud, J. M. (1995). Phylogenetic relationships and orthogenetic evolution of coat color among Tamarins (Genus *Saguinus*). *Systematic Biology* 44, 515–532.

Jamieson, C. J. (1998). Calanoid copepod biogeography in New Zealand. *Hydrobiologia* 367, 189–197.

Jamieson, G. M. (1981). Historical biogeography of Australian Oligochaeta. In *Ecological biogeography of Australia* (ed. A. Keast), pp. 887–921. Junk, The Hague.

Janke, A, Feldmaier-Fuchs, G., Thomas, W. K., von Haeseler, A., and Pääbo, S. (1994). The marsupial mitochondrial genome and the evolution of placental mammals. *Genetics* 137, 243–256.

Jeffrey, C. (1988). [Review of] The Plant Book. *Kew Bulletin* 43, 722–724.

Jell, P. A., and Duncan, P. M. (1986). Invertebrates, mailing insects, from the freshwater, Lower Cretaceous, Koonwarra fossil bed (Korunburra Group) South Gippsland, Victoria. *Memoirs of the Association of Australian Paleontology* 3, 111–205.

Jinks-Robertson, S., and Petes, T. D. (1993). Experimental determination of rates of concerted evolution. *Methods in Enzymology* 224, 631–646.

Johnson, K., and Balogh, G. (1977). Studies in the Lycaeninae (Lycaenidae) 2. Taxonomy and evolution of the nearctic *Lycaena rubidus* complex with descriptions of a new species. *Allyn Museum Bulletin 43*:1–62.

Jones, D. L. (1990). Synopsis of late Paleozoic and Mesozoic terrane accretion within the Cordillera of western North America. *Philosophical Transactions of the Royal Society of London Series A* 331, 479–486.

Jones, J. H. (1986). Evolution of the Fagaceae: the implications of foliar features. *Annals of the Missouri Botanical Garden* 73, 228–275.

Joseph, L., Moritz, C., and Hugall, A. (1995). Molecular support for vicariance as a source of diversity in rainforest. *Proceedings of the Royal Society of London Series B* 260, 177–182.

Juste, J., and Ibáñez, C. (1992). Taxonomic review of *Miniopterus minor* Peters, 1867 (Mammalia: Chiroptera) from western central Africa. *Bonner Zoologische Beiträge* 43, 355–365.

Kamp, P. J. J. (1991). Late Oligocene Pacific-wide tectonic event. *Terra Nova* 3, 65–69.

Kaneshiro, K. Y., Gillespie, R. G., and Carson, H. L. (1995). Chromosomes and male genitalia of Hawaiian *Drosophila*. In *Hawaiian biogeography: evolution of a hotspot* (eds. W. L. Wagner, and V. A. Funk), pp. 57–71. Smithsonian Institution Press, Washington, DC.

Katz, M. B. (1987). East African Rift and northeast lineaments: continental spreading-transform system? *Journal of African Earth Sciences* 6, 103–107.

Kavanaugh, D. H. (1980). Insects of western Canada, with special reference to certain Carabidae (Coleoptera): Present distribution patterns and their origins. *The Canadian Entomologist* 112, 1129–1144.

Keast, A. (1991). Panbiogeography: then and now. *The Quarterly Review of Biology* 66, 467–472.

Keay, R., and Stafleu, F. (1952). *Erismadelphus. Acta Botanica Neerlandica* 1, 594–99.

Kellman, M., and Tackaberry, R. (1993). Disturbance and tree species coexistence in tropical riparian forest fragments. *Global Ecology and Biogeography Letters* 3, 1–9.

Kellogg, J. N., and Vega, V. (1995). Tectonic development of Panama, Costa Rica, and the Colombian Andes: constraints from global positioning system geodetic studies and gravity. In *Geologic and tectonic development of the Caribbean plate boundary in southern Central America* (ed. P. Mann), pp. 75–90. Geological Society of America Special Paper 295, Boulder, Colorado.

Kimura, M. (1983). *The neutral theory of molecular evolution*. Cambridge University Press, Cambridge.

King, L. C. (1967). *South African scenery: a textbook of geomorphology*, 3rd ed. Oliver and Boyd, Edinburgh.

Kingdon, J. (1974). *East African mammals*, Vol. 2, Part A. Academic Press, New York.

Kiriakoff, S. G. (1954). Chorologie et systématique phylogénétique. *Bulletin Annales de la Societe Royale d'Entomologie de Belgique* 90, 185–198.

Kiriakoff, S. G. (1959). Systematics and typology. *Systematic Zoology* 8, 117–118.

Kiriakoff, S. G. (1961). Filosofische grondslagen van de biologische systematik. *Natuurwetenschappelijk Tijdschrift* 42, 35–57.

Kiriakoff, S. G. (1964). La vicariance géographique et la taxonomie. *Compte Rendu Sommaire des Séances Société de Biogéographie* 359–361, 103–115.

Kiriakoff, S. G. (1967). Biogeography and taxonomy. *Bulletin of the Indian National Science Academy* 34, 219–224.

Kiriakoff, S. G. (1981). La biogéographie du point vue de la cladistique. *Compte Rendu Sommaire des Séances Société de Biogéographie* 57, 51–62.

Klicka, J., and R. M. Zink. (1997). The importance of recent ice ages in speciation: a failed paradigm. *Science 277*, 1666—1669.

Knapp, S. (1991). A cladistic analysis of the *Solanum sessile* species groups, Section *Geminata pro parte* Solanaceae). *Botanical Journal of the Linnean Society* 106, 73–89.

Kolbek, J., and Alves, R. J. V. (1993). Some vicariating plant communities in Brazil, Malaysia and Singapore. *Vegetatio* 109, 15–27.

Koopman, K. F. (1984). Bats. In *Orders and families of recent mammals of the world* (eds. S. Anderson and J. Knox Jones Jr.), pp. 145–186. John Wiley and Sons, New York.

La Greca, M. (1990). The insect biogeography of west Mediterranean islands. *Atti dei Convegni Lincei* 85, 469–491.

Ladiges, P. Y., and C. J. Humphries. (1986). Relationships in the stringybarks, *Eucalyptus* L'Hérit, informal subgenus *Monocalyptus* series *Capitellatae* and *Olseniae*: phylogenetic hypotheses, biogeography and classification. *Australian Journal of Botany* 34, 603–632.

Laj, C., Mitouard, P., Roperch, P., Kissel, C., Mourier, T., and Megard, F. (1989). Paleomagnetic rotations in the coastal areas of Ecuador and Northern Peru. In *Paleomagnetic rotations and continental deformation* (eds. C. Kissel and C. Laj), pp. 489–511. Kluwer Academic Publishers, Dordrecht.

Lambert, R. St J., and Chamberlain, V. E. (1988). Cordilleria revisited, with a three-dimensional model for Cretaceous tectonics in British Columbia. *Journal of Geology* 96, 47–60.

Lanteri, A. A. (1992). Systematics, cladistics and biogeography of a new weevil genus, *Galapaganus* (Coleoptera: Curculionidae) from the Galápagos Islands, and coasts of Ecuador and Perú. *Transactions of the American Entomological Society* 118, 227–267.

Lanza, B. (1984). Sul significato biogeografico delle isole fossilo, con particulare riferimento all archipelago Pliocenico della Toscana. *Atti Societa Italiana di Science Naturali Museo Civicodi Storia Naturale di Milano* 125, 145–158.

Lapierre, H. J., Rouer, O., and Coulon, C. (1990). Model for the geodynamic evolution of the Western North American Cordillera during Paleozoic and early Mesozoic. In *Terrane analysis of China and the Pacific Rim* (eds. T. J. Wiley, D. G. Howell, and F. L. Wong), pp. 103–131. Circum-Pacific Council for Energy and Mineral Resources, Houston, Texas.

Larsson, S. G. (1978). Baltic amber—a palaeobiological study. *Entomonograph* 1, 1–192.

Lawes, M. J. (1990). The distribution of the samango monkey and forest history in southern Africa. *Journal of Biogeography* 17, 669–80.

Lebrun, J. (1947). *La végétation de la plaine alluviale au sud du Lac Edouard*. Institut des Parcs Nationaux du Congo Belge, Bruxelles.

Leenhouts, P. (1958). Connaraceae. *Flora Malesiana* I 5, 495–54I.

Leestmans, R. (1978). Problémes de spéciation dans le genre *Vanessa. Vanessa vulcana* Godart stat. nov. et *Vanessa buana* Frhst. stat. nov.: bonae species (Lepidoptera Nymphalidae). *Linneana Belgica* 7, 130–156.

Leis, J. M. (1984). Tetraodontiformes relationships. *American Society of Ichthyologists and Herpetologists Special Publication* 1, 459–463.

Letouzey, R. (1966). Etude phytogéographique du Cameroun. *Adansonia* 6, 205–15.

Leviton, A. E., and Anderson, S. C. (1984). Description of a new species of *Cyrtodactylus* from Afghanistan with remarks on the status of *Gymnodactylus cergipes* and *Cyrtodactylus fedschenkoi*. *Journal of Herpetology* 18, 270–276.

Levyns, M. R. (1964). Migrations and origins of the Cape flora. *Transactions of the Royal Society of South Africa* 37, 85–108.

Lewin, R. (1985). Hawaiian *Drosophila*: Young islands, old flies. *Science* 229, 1072–1074.

Lichtwardt, R. W. (1995). Biogeography and fungal systematics. *Canadian Journal of Botany* 73, 5731–5737.

Liebherr, J. K. (1986). Cladistic analysis of North American Platynini and revision of the *Agonum extensicolle* species group. *University of California Publications in Entomology* 106, 1–198.

Lindemann, D. (1990). Phylogeny and zoogeography of the New World terrestrial amphipods (landhoppers) (Crustacea: Amphipoda; Talitridae). *Canadian Journal of Zoology* 69, 1104–1116.

Linder, H. P. (1987). The evolutionary history of the Poales/Restionales—a hypothesis. *Kew Bulletin* 42, 279–318.

Linder, H. P., and Crisp, M. D. (1996). *Nothofagus* and Pacific biogeography. *Cladistics* 11, 5–32.

Lindsay, S. L. (1987). Geographic size and non-size variation in Rocky Mountain *Tamiasciurus hudsonicus*: significance in relation to Allen's rule and vicariant biogeography. *Journal of Mammalogy* 68, 39–48.

Londoño, A. C., Alvarez, E., Forero, E., and Morton, C. M. (1995). A new genus and species of Dipterocarpaceae from the Neotropics. I. Introduction, taxonomy, ecology, and distribution. *Brittonia* 47, 225–236.

Long, J. A. (1992). *Gogodipterus paddyensis* (Miles) *gen. nov.*, a new chirodipterid lungfish from the late Devonian Gogo formation, Western Australia. *Beagle* 9, 1–33.

Lonnberg, E. (1929). The development and distribution of the African fauna in connection with and depending upon climatic change. *Arkiv foer Zoologi* 21A, 1–33.

Lourenço, W. R., and Blanc, C. P. (1994). Biodiversité et biogéographie evolutive. *Biogeographica* 70, 49–57.

Lovis, J. D. (1990). Timing, exotic terranes, angiosperm diversification, and panbiogeography. *New Zealand Journal of Zoology* 16, 713–729.

Lynch, J. D. (1986). Origins of the high Andean herpetological fauna. In *High Altitude Tropical Biogeography*. (eds. F. Vuilleumier and M. Monasterio), pp. 478–499. Oxford University Press, New York and Oxford.

Lynch, J. D. (1989). The gauge of speciation: on the frequencies of modes of speciation. In *Speciation and its consequences* (eds. D. Otte and J. A. Endler), pp. 527–553. Sinauer Associates, Sunderland, Massachusetts.

MacFadden, B. J. (1980). Rafting mammals or drifting islands?: biogeography of the greater Antillean insectivores *Nesophontes* and *Solenodon*. *Journal of Biogeography* 7, 11–22.

Maldonado, M., and Uriz, M.-J. (1995). Biotic affinities in a transitional zone between the Atlantic and Mediterranean: a biogeographical approach based on sponges. *Journal of Biogeography* 22, 89–110.

Mallet, J. (1995). A species definition for the Modern Synthesis. *Trends in Ecology and Evolution* 10, 294–299.

Malusa, J. (1992). Phylogeny and biogeography of the pinyon pines (*Pinus* subsect. *Cembroides*). *Systematic Botany* 17, 42–66.

Manly, B. F. J. (1991). *Randomization and Monte Carlo methods in biology*. Chapman and Hall, London.

Mann, P. (1995). Preface. In *Geologic and tectonic development of the Caribbean plate boundary in southern Central America* (ed. P. Mann), pp. xi–xxxii. Geological Society of America Special Paper 295, Boulder, Colorado.

Margules, C. R., and Austin, M. P. (1994). Biological models for monitoring species decline: the construction and use of data bases. *Royal Society Philosophical Transactions. Biological Sciences* 344, 69–75.

Marshall, C. R. (1990). The fossil record and estimating divergence times between lineages: maximum divergence times and the importance of reliable phylogenies. *Journal of Molecular Evolution* 30, 400–408.

Martin, P. G., and Dowd, J. M. (1991). Application of evidence from molecular biology and the biogeography of angiosperms. *Australian Systematic Botany* 4, 111–116.

Martin, P. G., Gierl, A., and Saedler, H. (1989). Molecular evidence for pre-Cretaceous angiosperm origins. *Nature* 339, 46–48.

Martin, T. (1994). African origin of cavimorph rodents is indicated by incisor enamel microstructure. *Paleobiology* 20, 5–13.

Maruyama, S., Liou, J. G., and Seno, T. (1989). Mesozoic and Cenozoic evolution of Asia. In *The Evolution of the Pacific Ocean Margins* (ed. Z. Ben-Avraham), pp. 75–99. Oxford University Press, Oxford.

Marx, H., and Rabb, G. B. (1970). Character analysis: an empirical approach applied to advanced snakes. *Journal of Zoology (London)* 161, 525–548.

Marx, H., and Rabb, G. B. (1972). Phyletic analysis of fifty characters of advanced snakes. *Fieldiana (Zoology)* 63.

Mathis, W. N., and Wirth, W. W. (1978). A new genus near *Canaceoides* Cresson, three new species and notes on their classification (Diptera: Canacidae). *Proceedings of the Entomological Society of Washington* 80, 524–537.

Matile, L. (1982). Systématique, phylogénie et biogéographie des Diptères Keroplatidae des petites Antilles et de Trinidad. *Bulletin du Muséum National d'Histoire Naturelle. Paris 4è Série Section A 4*, 189–235.

Matile, L. (1990). Recherches sur la systématique et l'evolution des Keroplatidae (Diptera, Mycetophiloidea). *Mémoires du Muséum National d'Histoire Naturelle, Paris 4è Série Section A* 148, 1–682.

Matthews, C. (1990). Special issue on panbiogeography. *New Zealand Journal of Zoology* 16, i–iv, 471–808.

Matthews, J. V. (1977). Tertiary Coleoptera fossils from the North American Arctic. *The Coleopterists' Bulletin* 31, 297–308.

Mattson, P. H. (1984). Caribbean structural breaks and plate movements. *Geological Society of America Memoir* 162, 131–152.

Maxson, L., and Heyer, W. R. (1982). Leptodactylid frogs and the Brasilian

Shield: an old and continuing adaptive relationship. *Biotropica* 14, 10–15.

Maxson, L., and Roberts, J. D. (1984). Albumin and Australian frogs: molecular data a challenge to speciation model. *Science* 225, 957–958.

Mayden, R. L. (1985). Biogeography of Ouachita highland fishes. *Southwest Naturalist* 30, 195–211.

May, R. M. (1994). Conceptual aspects of the quantification of the extent of biological diversity. *Transactions of the Royal Society of London B* 345, 13–20.

Maynard Smith, J., and Vida, G. (eds.) (1990). *Organizational constraints on the dynamics of evolution.* Manchester University Press, Manchester, UK.

Mayo, S. J. (1993). Aspects of Aroid classification. In *The Africa-South America connection* (eds. W. George, and R. Lavocat), pp. 44–58. Clarendon Press, Oxford.

Mayr, E. (1942). *Systematics and the origin of species.* Columbia University Press, New York.

Mayr, E. (1982a). *The growth of biological thought*. Belknap Press, Cambridge, Massachusetts.

Mayr, E. (1982b). [Review of] Vicariance biogeography. *The Auk* 99, 618–620.

Mayr, E., and Ashlock, P. D. (1991). *Principles of systematic zoology*. 2nd ed. McGraw-Hill Inc., New York.

McAllister, D. E., Platania, S. P., Schueler, F. W., Baldwin, M. E., and Lee, D. S. (1986). Ichthyofaunal patterns on a geographic grid. In *The zoogeography of North American freshwater fishes* (eds. C. H. Hocutt and E. O. Wiley), pp. 17–51. John Wiley and Sons, New York.

McGuire, B., and Ashton, P. S. (1977). Pakaraimoideae, Dipterocarpaceae of the western hemisphere: Systematic, geographic and phyletic considerations. *Taxon* 26, 341–385.

McKenna, M. C. (1983). Holarctic landmass rearrangement, cosmic events and Cenozoic terrestrial organisms. *Annals of the Missouri Botanical Garden* 70, 459–490.

McLaughlin, S. P. (1994). Floristic plant geography: the classification of floristic areas and floristic elements. *Progress in Physical Geography* 18, 185–208.

Meacham, C. A. (1984). The role of hypothesized direction of characters in the estimation of evolutionary history. *Taxon* 33, 26–38.

Melville, R. (1982). The biogeography of *Nothofagus* and *Trigonobalanus* and the origin of the Fagaceae. *Botanical Journal of the Linnean Society of London* 85, 75–88.

Michaux, B. (1991). The evolution of the Ancillinae with special reference to New Zealand Tertiary and recent species of *Amalda* H. & A. Adams, 1853 (Gastropod: Olividae: Ancillinae). *Venus (Japanese Journal of Malacology)* 50, 130–149.

Michaux, B. (1994). Land movements and animal distributions in east Wallacea (eastern Indonesia, Papua New Guinea and Melanesia). *Palaeogeography, Palaeoclimatology, Palaeoecology* 112, 323–343.

Mickleburgh, S. P., Hutson, A. M., and Racey, P. A. (1992). *Old World bats: an action plan for their conservation.* International Union for Conservation of Nature, Gland, Switzerland.

Millar, C. I. (1993). Impact of the Eocene on the evolution of *Pinus* L. *Annals of the Missouri Botanical Garden* 80, 471–498.
Miller, A. I., and Mao, S. (1995). Association of orogenic activity with the Ordovician radiation of marine life. *Geology* 23, 305–308.
Miller, L. D. (1987). A new subspecies of *Heraclides aristodemus* from Crooked Island, Bahamas, with a discussion of the distribution of the species. *Bulletin of the Allyn Museum* 113.
Miller, L. D., and Miller, J. Y. (1990). Nearctic *Aglais* and *Nymphalis* (Lepidoptera, Nymphalidae): Laurasia revisited? *The Entomologist* 109, 106–115.
Miller, R. I. (1994). Setting the scene. In *Mapping the diversity of nature* (ed. R. I. Miller), pp. 1–17. Chapman and Hall, London.
Miller, W. B., and Naranjo-Garcia, E. (1991). Familial relationships and biogeography in the western American and Caribbean Helicoidea (Mollusca: Gastropoda: Pulmonata). *American Malacological Bulletin* 8, 147–153.
Milner, A. R., and Norman, D. B. (1984). The biogeography of advanced ornithopod dinosaurs (Archosauria: Ornithischia)—a cladistic-vicariance model. In *Third symposium on Mesozoic terrestrial ecosystems, short papers* (eds. W. E. Reif and F. Westphal), pp. 145–150. ATTEMPTO Verlag, Tübingen.
Minckley, W. L., Hendrickson, D. A., and Bond, C. E. (1986). Geography of western North American freshwater fishes: description and relationships to intracontinental tectonism. In *The zoogeography of North American freshwater fishes* (eds. C. H. Hocutt, and E. O. Wiley), pp. 519–613. John Wiley and Sons, New York.
Mlot, C. (1995). In Hawaii, taking inventory of a biological hot spot. *Science* 269, 322–323.
Molnar, R. E. (1997). Biogeography for dinosaurs. In *The Complete Dinosaur* (eds. J. O. Farlow and M. K. Brett-Surman), pp. 581–606. Indiana University Press, Bloomington.
Monger, J. W. H. (1993). Canadian Cordilleran tectonics: from geosynclines to crystal collage. *Canadian Journal of Earth Sciences* 30, 209–231.
Montgomery, H., Pessagno, E. A., Lewis, J. F., and Schellekens, J. (1994). Paleogeography of Jurassic fragments in the Caribbean. *Tectonics* 13, 725–732,
Morain, S. A. (1984). *Systematic and regional biogeography.* Van Nostrand Reinhold Company. New York.
Moritz, C. (1995). Use of molecular phylogenies for conservation. *Philosophical Transactions of the Royal Society of London* B. 349, 113–118.
Morrone, J. J. (1992). Revisión sistemática análisis cladistico y biogeografía histórica de los géneros *Falklandius* Enderlein y *Lanteriella* Gen. nov. (Coleoptera: Curculionidae). *Acta Entomológica Chilena* 17, 157–174.
Morrone, J. J. (1993). Revisión sistemática de un nuevo género de rhytirrhinini (Coleoptera, Curculionidae), con un análisis biogeográfico del dominio subantárctico. *Boletín de la Sociedad Biología de Concepción Chile* 64, 121–145.
Morrone, J. J. (1994). Distributional patterns of species of Rhytirrhinini (Coleoptera: Curculionidae) and the historical relationships of the Andean provinces. *Global Ecology and Biogeography Letters* 4, 188–194.

Morrone, J. J. (1996). Austral biogeography and relict weevil taxa (Coleoptera: Nemonychidae, Belidae, Brentidae, and Caridae). *Journal of Comparative Biology* 1, 123–127.

Morrone, J. J., and Crisci, J. V. (1992). Aplicación de metodos filogenéticos y panbiogeográficos en la conservacíon de la diversidad biológica. *Evolución Biológica* 6, 53–66.

Morrone, J. J., and Crisci, J. V. (1995). Historical biogeography: introduction to methods. *Annual Review of Ecology and Systematics* 26, 373–401.

Morrone, J. J., Espinosa-Organista, D., Llorente-Bousquets, J. L. (1996). *Manual de biogeografía histórica.* Universidad Nacional Autónoma de México.

Morrone, J. J., and Lopretto, E. C. (1994). Distributional patterns of freshwater Decapoda (Crustacea: Malacostraca) in southern South America: a panbiogeographic approach. *Journal of Biogeography* 21, 97–109.

Mortimer, N. (1993). Geology of the Otago schist and adjacent rocks. Scale 1:500 000. *Institute of Geological and Nuclear Sciences Map 7*. 1 Sheet. Institute of Geological and Nuclear Sciences Ltd., Lower Hutt, New Zealand.

Mourier, T., Laj, C., Megard, F., Roperch, P., Mitouard, P., and Farfan Medrano, A. (1988). An accreted continental terrane in northwestern Peru. *Earth and Planetary Science Letters* 88, 182–192.

Moya, A., Galiana, A., and Ayala, F. J. (1995). Founder-effect speciation theory: failure of experimental corroboration. *Proceedings of the National Academy of Sciences* USA 92, 3983–3986.

Murphy, R. (1975). Two new blind snakes (Serpentes: Leptotyphlopidae) from Baja California, Mexico, with a contribution to the biogeography of peninsular and insular herpetofauna. *Proceedings of the California Academy of Science* 40, 93–107.

Murphy, R. (1983a). Paleobiogeography and patterns of genetic differentiation of the Baja California herpetofauna. *California Academy of Science Occasional Papers* 137, 1–48.

Murphy, R. (1983b). The reptiles: origins and evolution. In *Island biogeography in the Sea of Cortez* (eds. T. J. Case and M. L. Cody), pp. 130–158. University of California Press, Berkeley.

Musser, G. G., and Holden, M. E. (1991). Sulawesi rodents (Muridae: Murinae): morphological and geographical boundaries of species in the *Rattus hoffmani* groups and a new species from Pulau Penang. *Bulletin of the American Museum of Natural History* 206, 322–413.

Myers, A. A. (1988). A cladistic and biogeographic analysis of the Aorinae sub-family nov. *Crustacea Supplement* 13, 167–192.

Myers, A. A. (1990). [Review of] Panbiogeography. *Journal of Natural History* 24, 1333–34.

Myers, A. A. (1991). How did Hawaii accumulate its biota? A test from the Amphipoda. *Global Ecology and Biogeography Letters* 1, 24–29.

Myers, A. A. (1994). Biogeographic patterns in marine systems and the controlling processes at different scales. In *Aquatic ecology: scale, pattern and process* (eds. P. S. Giller, A. G. Hildrew, and D. G. Rafaelli), pp. 547–574. Blackwell Scientific Publications, Oxford.

Myers, A. A., and Giller, P. S. (1990). *Analytical biogeography.* Chapman and Hall, London.

Myers, G. (1990). Every picture tells a story: illustrations. In *Representation in scientific practice* (eds. M. Lynch and S. Woolgar), pp. 231–256. The MIT Press, Cambridge.

Naylor, B. G., and Fox, R. C. (1993). A new ambystomatid salamander *Dicamptodon antiquus* n. sp., from the Paleocene of Alberta, Canada. *Canadian Journal of Earth Science* 30, 814–818.

Nelson, G. (1973). Comments on Leon Croizat's biogeography. *Systematic Zoology* 22, 312–320.

Nelson, G. (1978). From Candolle to Croizat: comments on the history of biogeography. *Journal of the History of Biology* 11, 269–305.

Nelson, G. (1983). Vicariance and cladistics: historical perspectives with implications for the future. In *Evolution, time, and space: the emergence of the biosphere* (eds. R. W. Sims J. H. Price, and P. E. S. Whalley), pp. 469–492. Academic Press, London.

Nelson, G. (1984). Identity of the anchovy *Hildebrandichthys setiger* with notes on relationships and biogeography of the genera *Engraulis* and *Cetengraulis*. *Copeia* 7, 477–427.

Nelson, G. (1985). A decade of challenge: the future of biogeography. *Journal of History of Earth Sciences Society* 4, 187–196.

Nelson, G. (1989). Cladistics and evolutionary models. *Cladistics* 5, 275–289.

Nelson, G., and Ladiges, P. (1990). Biodiversity and biogeography. *Journal of Biogeography* 17, 559–560.

Nelson, G., and Platnick, N. (1981). *Systematics and biogeography: cladistics and vicariance*. Columbia University Press, New York.

Nelson, G., and Rosen, D. E. (1981). *Vicariance biogeography: a critique*. Columbia University Press, New York.

New, T. R. (1995). Butterfly conservation in Australasia—an emerging awareness and an increasing need. In *Ecology and conservation of butterflies* (ed. A. S. Pullin), pp. 304–315. Chapman and Hall, London.

Newbigin, M. I. (1912). *Man and his conquest of nature*. Adams and Charles Black, London.

Nixon, K. C., and Crepet, W. L. (1989). *Trigonobalanus* (Fagaceae): taxonomic status and phylogenetic relationships. *American Journal of Botany* 76, 828–841.

Noonan, G. R. (1989). Biogeography of North American and Mexican Insects, and a critique of vicariance biogeography. *Systematic Zoology* 37, 366–384.

Nooteboom, H. (1962). Simaroubaceae. *Flora malesiana* 1, 193–226.

Nordenstam, B. (1968). Phytogeography of the genus *Euryops*. *Opera Botanica* 23.

Notenboom, J. (1991). Marine regressions and the evolution of groundwater dwelling amphipods (Crustacea). *Journal of Biogeography* 18, 437–454.

O'Brien, S. J., and Mayr, E. (1991). Bureaucratic mischief: recognizing endangered species and subspecies. *Science* 251, 1187–1188.

Ohno, S. (1970). *Evolution by gene duplication*. Springer-Verlag, Berlin.

Olsen, J. S. (1979). Systematics of *Zaluzania* (Asteraceae: Heliantheae). *Rhodora* 81, 449–501.

Olson, S., and Ames, P. L. (1984). *Promerops* as a thrush and its implications for the evolution of nectarivory in birds. *Ostrich* 56, 213–218.

Oosterbroek, P., and Arntzen, J. (1992). Area-cladograms of circum-Mediterranean taxa in relation to Mediterranean palaeogeography. *Journal of Biogeography* 19, 3–20.

Oreskes, N. (1988). The rejection of continental drift. *Historical Studies in the Physical and Biological Sciences* 18, 311–348.

Oyama, S. (1985). *The ontogeny of information: developmental systems and evolution*. Cambridge University Press, Cambridge.

Page, R. D. M. (1987). Graphs and generalized tracks: quantifying Croizat's panbiogeography. *Systematic Zoology* 36, 1–17.

Page, R. D. M. (1988). Quantitative cladistic biogeography: constructing and comparing area cladograms. *Systematic Zoology* 37, 254–270.

Page, R. D. M. (1990a). New Zealand and the new biogeography. *New Zealand Journal of Zoology* 16, 471–483.

Page, R. D. M. (1990b). Tracks and trees in the Antipodes. *Systematic Zoology* 39, 288–299.

Page, R. D. M. (1991). Random dendrograms and null hypotheses in cladistic biogeography. *Systematic Zoology* 40, 54–62.

Pagel, M. D., and Harvey, P. H. (1988). Recent developments in the analysis of comparative data. *The Quarterly Review of Biology* 37, 254–270.

Pagel, M. D., and Harvey, P. H. (1991). *The comparative method in evolutionary biology*. Oxford University Press, Oxford.

Pandolfi, J. M. (1992). Successive isolation rather than evolutionary centres for the origination of Indo-Pacific reef corals. *Journal of Biogeography* 19, 593–609.

Parenti, L. R. (1991). Ocean basins and the biogeography of freshwater fishes. *Australian Systematic Botany* 4: 137–149.

Patrick, B. H. (1984). Lammermoor-Lammerlaw—A tussockland national reserve in Eastern Otago? *Forest and Bird* 15, 7.

Patrick, B. H. (1990). Panbiogeography and the amateur naturalist with reference to conservation implications. *New Zealand Journal of Zoology* 16, 749–755.

Patrick, B. H. (1991). Description of a new species of Crambidae (Crambinae: Lepidoptera) from New Zealand. *New Zealand Journal of Zoology* 18, 357–362.

Patrick, B. H., Barratt, B. I. P., Ward, J. B., and McLellan, I. D. (1993). *Insects of the Waipori ecological district: Lammerlaw ecological Region*. Otago Conservancy Miscellaneous Series No. 16. Department of Conservation, Dunedin.

Patterson, B. D. (1980). Montane mammalian biogeography in New Mexico. *The Southwestern Naturalist* 25, 33–40.

Patterson, B. D. (1982). Pleistocene vicariance, montane islands and the evolutionary divergence of some chipmuncks (genus *Eutamias*). *Journal of Mammalogy* 63, 387–398.

Patterson, C. (1981). Methods of paleobiogeography. In *Vicariance biogeography: a critique* (eds. G. Nelson and D. E. Rosen), pp. 446–489. Columbia University Press, New York.

Peiyu, Y., and Stork, N. E. (1991). New evidence on the phylogeny and biogeography of the Amphizoidae: discovery of a new species from China (Coleoptera). *Systematic Entomology* 16, 253–256.

Pennington, T. D. (1981). *Meliaceae.* New York Botanical Garden, New York.

Perkins, P. D. (1980). Aquatic beetles of the family Hydraenidae in the Western Hemisphere: classification, biogeography and inferred phylogeny (Insecta: Coleoptera). *Quaestiones Entomologicae* 16, 5–554.

Petuch, E. J. (1981). A relict neogene caenogastropod fauna from northern South America. *Malacologia* 20, 307–347.

Phipps, J. B. (1975). Kilometric distance. *Canadian Journal of Botany* 53, 116–1119.

Phipps, J. B. (1983). Biogeographic, taxonomic, and cladistic relationships between east Asiatic and North American *Crategus. Annals of the Missouri Botanical Garden* 70, 667–700.

Pindell, J. L., and Barrett, S. F. (1991). Geological evolution of the Caribbean region; a plate-tectonic perspective. In *The Caribbean region* (eds. G. Dengo, and J. E. Case), pp. 405–432. The Geological Society of America, Boulder, Colorado.

Pitkin, L. D. (1988). The Holarctic genus *Teleispsis*: host-plants, biogeography, and cladistics (Lepidoptera: Gelechiidae). *Entomologica Scandinavica* 19, 143–191.

Platnick, N. (1976). Drifting spiders or continents? Vicariance biogeography of the spider subfamily Laroniinae (Araneae: Gnaphosidae). *Systematic Zoology* 25, 101–109.

Platnick, N. I. (1992). Patterns of biodiversity. In *Systematics, ecology, and the biodiversity crisis* (ed. N. Eldredge), pp. 15–24. Columbia University Press, New York.

Platnick, N., and Nelson, G. (1984). Composite areas in vicariance biogeography. *Systematic Zoology* 33, 328–335.

Platnick, N., and Nelson, G. (1989). Spanning-tree biogeography: shortcut, detour, or dead-end? *Systematic Zoology* 37, 410–419.

Poinar, G. O. Jr., and Cannatella, D. C. (1987). An upper Eocene frog from the Dominican Republic and its implication for Caribbean biogeography. *Science* 237, 1215–1216.

Polhemus, D. A. (1996). Island arcs, and their influence on Indo-Pacific biogeography. In *The origin and evolution of Pacific Island biotas, New Guinea to Eastern Polynesia: patterns and processes* (eds. A. Keast and S. E. Miller), pp. 51–66. SPB Academic Publishing, Amsterdam.

Popoff, M. (1988). Du Gondwana à l'Atlantique sud: les connexions du fosse de la Benoue avec les bassins du Nord-Est brésilien jusqu'à l'ouverture du golfe de Guinée au Cretace inférieur. *Journal of the African Earth Sciences* 7, 409–431.

Popper, K. (1972). *Conjectures and refutations: the growth of scientific knowledge.* Routledge and Kegan Paul, London.

Powell, J. R., and DeSalle, R. (1995). *Drosophila* molecular phylogenies and their uses. *Evolutionary Biology* 28, 87–138.

Poynton, J. C. (1964). Southern African Amphibia. *Annals of the Natal Museum.* 17, 1–334.

Poynton, J. C. (1983). The dispersal versus vicariance debate in biogeography. *Bothalia* 14, 455–460.

Pressey, R. L., Humphries, C. J., Margules, C. R., Vane-Wright, R. I. and Williams, P. H. (1993). Beyond opportunism: key principles for systematic reserve selection. *Trends in Ecology and Evolution* 8, 124–128.

Pullin, A. S., ed. (1995). *Ecology and conservation of butterflies.* Chapman and Hall, London.

Radford, A. (1986). *Fundamentals of plant systematics.* Harper and Row, New York.

Rainboth, W. J. (1996). The taxonomy, systematics, and zoogeography of *Hypsibarbus*, a new genus of large barbs (Pisces, Cyprinidae) from the rivers of Southeastern Asia. *University of California Publications in Zoology 129.*

Rapoport, E. H. (1982). *Areography: geographical strategies of species* [trans. B. Drausal]. Pergamon Press, Oxford.

Raven, P. H., and Wilson, E. O. (1992). A fifty-year plan for biodiversity surveys. *Science* 258, 1099–1100.

Rehder, H. A (1980). The marine molluscs of Easter-island (Isla de Pascua) and Sala y Gomez. *Smithsonian Contributions to Zoology* 289, 1–167.

Renner, S. S., and Ricklefs, R. E. (1994). Systematics and biodiversity. *Trends in Ecology and Evolution* 9, 78.

Rickart, E. A., and Heaney, L. R. (1991). A new species of *Chrotomys* (Rodentia: Muridae) from Luzon Island, Philippines. *Proceedings of the Biological Society of Washington* 104, 387–398.

Ricklefs, R. E., and Schulter, D. (eds.) (1994). *Species Diversity in Ecological Communities.* University of Chicago Press, Chicago.

Riddle, B. R. (1995). Molecular biogeography of the pocket mice (*Perognathus* and *Chaetodipus*) and grasshopper mice (*Onchomys*): the Late Cenozoic development of a North American aridlands rodent guild. *Journal of Mammology* 76, 283–301.

Riddle, B. R., and Honeycutt, R. L. (1990). Historical biogeography in North American arid regions: an approach using mitochondrial-DNA phylogeny in grasshopper mice (genus *Onychomys*). *Evolution* 44, 1–15.

Rodriguez, G. (1986). Centers of radiation of freshwater crabs in the neotropics. In *Crustacean Biogeography* (eds. R. H. Gore and K. L. Heck), pp. 51–67. A. A. Balkema, Rotterdam.

Rodriguez, G. (1992). The freshwater crabs of America. *Editions de l'Orstom. Collection Faune Tropicale* 31, 1–189.

Rogers, G. M. (1989). The nature of the lower North Island floristic gap. *New Zealand Journal of Botany* 27, 221–241.

Rosa, D. (1918). *Ologenesi.* R. Bemporad and Figlio, Firenze.

Roscoe, D. J. (1990). Panbiogeography and the public. *New Zealand Journal of Zoology* 16, 757–761.

Rosen, B. R. (1984). Reef coral biogeography and climate through the late Cainozoic: just islands in the sun or a critical pattern of islands? In *Fossils and climate* (ed. P. J. Brenchley), pp. 201–262. Wiley, Chichester, UK.

Rosen, B. R. (1988). From fossils to earth history: applied historical biogeography. In *Analytical Biogeography* (eds. A. A. Myers and P. S. Giller), pp. 437–481. Chapman and Hall, London.

Rosen, B. R., and Smith, A. B. (1988). Tectonics from fossils? In *Gondwana and Tethys* (eds. M. G. Audley-Charles and A. Hallam), pp. 275–366. Oxford University Press, Oxford.

Rosen, D. E. (1974). Phylogeny and zoogeography of Salmoniform fishes and relationships of *Lepidogalaxias salamandroides*. *Bulletin of the American Museum of Natural History* 153, 269–325.

Rosen, D. E. (1976). A vicariance model of Caribbean biogeography. *Systematic Zoology* 24, 431–464.

Rosen, D. E. (1978). Vicariant patterns and historical explanation in biogeography. *Systematic Zoology* 27, 159–188.

Rosen, D. E. (1985). Geological hierarchies and biogeographic congruence in the Caribbean. *Annals of the Missouri Botanical Garden* 72, 636–659.

Rosenblatt, R. H., and Waples, R. S. (1986). A genetic comparison of allopatric populations of shore fish species from the eastern and central Pacific Ocean: dispersal or vicariance? *Copeia* 1986, 275–284.

Ross, C. A., and Ross, J. R. P. (1985). Carboniferous and early Permian biogeography. *Geology* 13, 27–30.

Ross, H. H. (1974). *Biological systematics*. Addison-Wesley, Reading, Massachusetts.

Rotondo, G. M., Springer, V. G., Scott, G. A. J., and Schlanger, S. O. (1981). Plate movement and island integration—a possible mechanism in the formation of endemic biotas, with special reference to the Hawaiian Islands. *Systematic Zoology* 30, 12–21.

Roughgarden, J. (1995). *Anolis lizards of the Caribbean: ecology, evolution, and plate tectonics*. Oxford University Press, New York.

Roure, F., and Sosson, M. (1986). Late Jurassic collision between a composite exotic block and the North American continent: a model for the Cordilleria building. *Bulletin de la Société Géologique de France Huitième Série 2*, 6, 945–959.

Rousseau, D. D., and Puisségur, J. J. (1990). Phylogenèse et biogéographie de *Retinella* (*Lyrodiscus*) Pilsbury (Gastropoda: Zonitidae). *Geobios* 23, 57–70.

Rowe, F. W. E. (1985). Six new species of *Asterodiscus* A. M. Clark (Echinodermata, Asteroidea), with a discussion of the origin and distribution of the Asterodiscididae and other 'amphi-Pacific' echinoderms. *Bulletin du Muséum National d'Histoire Naturelle. Section A: Zoologie Biologie et Ecologie Animales* 7, 531–577.

Ruvolo, M. (1994). Molecular evolutionary processes and conflicting gene trees: the hominid case. *American Journal of Physical Anthropology* 94, 89–113.

Saether, O. A. (1981). Orthocladiinae (Chironomidae: Diptera) from the British West Indies with descriptions of *Antillocladius* n.gen., *Lipurometriocnemus* n.gen., *Compterosmittia* n.gen., and *Diplosmittia* n.gen. *Entomologica Scandinavica Supplementum* 16, 1–46.

Salick, J., and Pong, P. Y. (1984). An analysis of termite faunae in Malayan rainforests. *Journal of Applied Ecology* 21, 547–561.

Sang, T., Crawford, D. J., and Stuessy, T. F. (1995). Documentation of reticulate evolution in peonies (*Paeonia*) using internal transcribed spacer sequences of nuclear ribosomal DNA: implications for biogeography and concerted evolution. *Proceedings of the National Academy of Science USA* 92, 6813–6817.

Sarich, V. M., and Cronin, J. E. (1980). South American mammal molecular systematics, evolutionary clocks, and continental drift. In *Evolutionary biology of the New World monkeys and continental drift* (eds. R. L. Ciochon and A. B. Chiarelli), pp. 399–421. Plenum Press, New York.

Saunders, J. H. (1961). *The wild species of Gossypium and their evolutionary history*. Oxford University Press, London.

Savage, J. M. (1982). The enigma of the central American herpetofauna: dispersals or vicariance? *Annals of the Missouri Botanical Garden* 69, 464–547.

Savard, L., Li, P., Strauss, S. H., Chase, M. W., Michaud, M., and Bousquet, J. (1994). Chloroplast and nuclear gene sequences indicate Late Pennsylvanian time for the last common ancestor of extant seed plants. *Proceedings of the National Academy of Science USA* 91, 5163–5167.

Schilder, F. A. (1952). *Einführung in die biotaxonomie (Formenkreislehre). Die entstehung der Arten durch räumliche sonderung*. Gustav Fischer, Jena.

Schimper, A. F. W. (1898). *Pflanzengeographie auf physiologischen grundlage*. Gustav Fischer, Jena.

Schindewolf, O. H. (1993). *Basic questions in paleontology: geological time, organic evolution and biological systematics*. University of Chicago Press, Chicago.

Schlanger, S. O., and Gillett, G. W. (1976). A geological perspective of the upland biota of Laysan atoll (Hawaiian Islands). *Biological Journal of the Linnean Society* 8, 205–216.

Schoener, A., and Schoener, T. W. (1984). Experiments on dispersal: short-term floatation of insular anoles with a review of similar abilities in other terrestrial animals. *Oecologia (Berlin)* 63, 289–294.

Schuster, R. M. (1976). Plate tectonics and its bearing on the geographical origin and dispersal of angiosperms. In *Origin and early evolution of angiosperms* (ed. C. B. Beck), pp. 48–138. Columbia University Press, New York.

Sclater, P. L. (1858). On the general geographical distribution of the members of the class Aves. *Journal of the Linnean Society (Zoology)* 2, 130–145.

Scott, M. J., Csuti, B., Smith, K., Estes, J. E., and Caicco, S. (1991). Gap analysis of species richness and vegetation cover: an integrated biodiversity conservation strategy. In *Balancing on the edge of extinction: the Endangered Species Act and lessons for the future* (ed. K. Kohm), pp. 282–297. Island Press, Washington, DC.

Scott, M. J., Davis, F., Csuti, B., Noss, R., Butterfield, B., Groves, C., Anderson, H., Caicco, S., D'Erchia, F., Edwards Jr., T. C., Ulliman, J., and Wright, R. G. (1993). Gap analysis: a geographic approach to protection of biological diversity. *Wildlife Monographs* 123, 1–41.

Shanahan, T. (1992). Selection, drift, and the aims of evolutionary theory. In *Trees of Life: Essays in Philosophy of Biology* (ed. P. Griffiths) pp. 133–161. Kluwer Academic Publishers, Dordrect.

Shaw, J. (1985). The relevance of ecology to species concepts in bryophytes. *The Bryologist* 88, 199–206.

Shear, W. A., Schawaller, W., and Bonamo, P. M. (1989). Record of Palaeozoic pseudoscorpions. *Nature* 341, 527–529.

Sibley, C. G., and Ahlquist, J. E. (1974). The relationships of the African sugarbirds (*Promerops*). *Ostrich* 45, 22–30.

Sibley, C. G., and Monroe, B. L. (1990). *Distribution and taxonomy of birds of the world*. Yale University Press, New Haven, Connecticut.

Siegel-Causey, D. (1991). Systematics and biogeography of North Pacific shags, with a description of a new species. *Occasional Papers of the Museum of Natural History, University of Kansas*. 140.

Simpson, G. G. (1945). The principles of classification and a classification of the mammals. *Bulletin of the American Museum of Natural History* 85, 1–350.

Simpson, G. G. (1961). *Principles of animal taxonomy.* Columbia University Press, New York.

Sims, R. W. (1980). A classification and the distribution of earthworms, suborder Lumbricina (Haplotaxida: Oligochaeta). *Bulletin of the British Museum Natural History (Zoology)* 39, 103–124.

Sites, J. W. Jr., Davis, S. K., Guerra, T, Iverson, J. B., and Snell, H. L. (1996). Character congruence and phylogenetic signal in molecular and morphological data sets: a case study in the living iguanas (Squamata, Iguanidae). *Journal of molecular biology and evolution* 13, 1087–1105.

Sippman, H. J. M. (1983). A monograph of the lichen family Megalosporaceae. *Biblioteca Lichenologica* 18.

Skottsberg, C. (1936). Antarctic plants in Polynesia. In *Essays in geobotany in honor of William Albert Setchell* (ed. T. H. Goodspeed), pp. 291–311. University of California Press, Berkeley.

Skottsberg, C. (1956). Derivation of the flora and fauna of Juan Fernandez and Easter Island. In *The National History of Juan Fernandez and Easter Island I* (ed. C. Skottsberg), pp. 193–438. Almquist and Wiksell, Uppsala.

Sluys, R. (1994). Explanations for biogeographic tracks across the Pacific Ocean: a challenge for paleogeography and historical biogeography. *Progress in Physical Geography* 18, 42–58.

Sluys, R. (1995). Platyhelminths as paleogeographical indicators. *Hydrobiologia* 305, 49–53.

Smith, A. B. (1992). Echinoid distribution in the Cenomanian: an analytical study in biogeography. *Palaeogeography, Palaeoclimatology, Palaeoecology* 92, 263–276.

Smith, C. H. (1990). Historical biogeography: geography as evolution, evolution as geography. *New Zealand Journal of Zoology* 16, 773–785.

Smith, L. B. (1934). Geographical evidence on the lines of evolution in the Bromeliaceae. *Botanische Jahrbuecher für Systematick* 66, 446–468.

Sneath, P. H. A. (1967). Conifer distributions and continental drift. *Nature 215*, 467–470.

Sober, E. (1988). *Reconstructing the past: parsimony, evolution and inference.* MIT Press, Cambridge, Massachusetts.

Southey, I. (1990). Biogeography of New Zealand's terrestrial vertebrates. *New Zealand Journal of Zoology* 16, 652–653.

Spanton, T. G. (1994). Vicariance events in the biogeographical history of *Panscopus* Schönherr weevils in North America (Coleoptera: Curculionidae: Entiminae). *The Canadian Entomologist* 126, 457–470.

Spence, J. R. (1982). Taxonomic status, relationships, and biogeography of *Anatrichis* Leconte and *Oodinus* Motschulsky (Carabidae: Oodini). *Coleopterists' Bulletin* 36, 567–580.

Springer, M. S. (1995). Molecular clocks and the incompleteness of the fossil record. *Journal of Molecular Evolution* 41, 531–538.

Springer, V. G. (1982). Pacific plate biogeography with special reference to shorefishes. *Smithsonian Contributions to Zoology* 367, 1–182.

Stace, C. A. (1989). *Plant taxonomy and biosystematics*, 2nd ed. Edwin Arnold, London.

Stafleu, F. (1954). A monograph of the Vochysiaceae. IV. *Erisma*. *Acta Botanica Neerlandica* 3, 458–80.

Stagman, J. G. (1978). An outline of the geology of Rhodesia. *Rhodesia Geological Survey Bulletin* 80.

Stearly, R. F. (1992). Historical ecology of the Salmoninae, with special reference to *Oncorhynchus*. In *Systematics, historical ecology, and North American freshwater fishes* (ed. R. L. Hayden), pp. 622–658. Stanford University Press, Stanford, California.

Stepien, C. A. (1992). Evolution and biogeography of the Clinidae (Teleostei: Blennioidei). *Copeia* 2, 375–392.

Stock, C. A. (1977). The taxonomy and zoogeography of the hadziid Amphipoda with emphasis on the West Indian taxa. *Studies on the fauna of Curacao and other Caribbean islands* 55, 1–81.

Stock, C. A. (1991). Insular groundwater biotas in the (sub)tropical Atlantic: a biogeographic synthesis. *Academia Nazionale dei Lincei Atti Convegni Lincei* 85, 695–713.

Stock, J. H. (1993). Some remarkable distribution patterns in stygobiont Amphipoda. *Journal of Natural History* 27, 807–819.

Stockhert, B., Maresch, W. V., Bix, M., Kaiser, C., Toetz, A., Kluge, R., and Kruckhans-Lueder, G. K. (1995). Crustal history of Margarita Island (Venezuela) in detail: Constraint on the Caribbean plate-tectonic scenario. *Geology* 23, 787–790.

Stoddart, D. R. (1985). *On geography and its history*. Blackwell Scientific Publications, Oxford.

Stoddart, D. R. (1992). Biogeography of the tropical Pacific. *Pacific Science* 46, 276–293.

Storey, B. C. (1995). The role of mantle plumes in continental breakup: case histories from Gondwanaland. *Nature* 377, 301–308.

Stott, P. (1981). *Historical plant geography*. George Allen and Unwin, London.

Stott, P. (1984). History of biogeography. In *Themes in biogeography* (ed. J. A. Taylor), pp. 1–24. Croom Helm, London.

Stuessy, T. F. (1979). Cladistics of *Melampodium* (Compositae). *Taxon* 28, 179–195.

Stuessy, T. F. (1990). *Plant taxonomy: the systematic evaluation of comparative data*. Columbia University Press, New York.

Stuessy, T. F., and Crisci, J. V. (1984). Problems in the determination of evolutionary directionality of character state change for phylogenetic reconstruction. In *Cladistics: perspectives on the reconstruction of evolutionary history* (eds. T. Duncan and T. F. Stuessy), pp. 71–87. Columbia University Press, New York.

Szymura, J. M. (1983). Genetic differentiation between hybridizing species *Bombina bombina* and *Bombina variegata* (Salientia, Discoglossidae) in Poland. *Amphibia-Reptilia* 4, 137–145.

Tambussi, C., Noriega, J., Gazdzicki, A., Tatur, A., Reguero, M. A., and Vizcaino, S. F. (1994). Ratite bird from the Paleogene of la Meseta formation, Seymour Island, Antarctica. *Polish Polar Research* 15, 15–20.

Tanai, T. (1986). Phytogeographic and phylogenetic history of the genus *Nothofagus* Bl. (Fagaceae) in the Southern Hemisphere. *Journal of the Faculty of Science, Hokkaido University, Series IV, 21 (4)*: 505–582.

Tanai, T. (1990). Euphorbiaceae and Icacinaceae from the Paleogene of Hokkaido Japan. *Bulletin of the National Science Museum Series C* 16, 91–118.

Tangelder, I. R. M. (1988). The biogeography of the Holarctic *Nephrotoma dorsalis* species-group (Diptera: Tipulidae). *Beaufortia* 38, 1–35.

Taylor, D. W. (1960). Distribution of the freshwater clam *Pisidium ultramontanum*: a zoogeographic inquiry. *American Journal of Science (A)* 258, 325–334.

Taylor, D. W. (1966). Summary of North American blancan non-marine mollusks. *Malacologia* 4, 1–172.

Taylor, D. W. (1988). Aspects of freshwater mollusc ecological biogeography. *Palaeogeography, Palaeoclimatology, Palaeoecology* 62, 511–576.

Taylor, D. W. (1990). Palaeogeographic relationships of angiosperms from the Cretaceous and early Tertiary of the North America area. *The Botanical Review* 56, 279–415.

Taylor, D. W., and Bright, R. C. (1988). Drainage history of the Bonneville Basin. *Utah Geological Association Publication* 16, 239–256.

Tehler, A. (1983). The genera *Dirina* and *Roccellina* (Roccellaceae). *Opera Botanica* 70, 1–86.

Terent'ev, P. V. (1965). *Herpetology*. Israel Program for Scientific Translations, Jerusalem.

Thewissen, J. G. M. (1992). Temporal data in phylogenetic systematics: an example from the Mammalian fossil record. *Journal of Palaeontology* 66, 1–8.

Thorne, R. F. (1975). Angiosperm phylogeny and geography. *Annals of the Missouri Botanical Garden* 62, 362–367.

Thorne, R. F. (1989). Phylogeny and phytogeography. *Rhodora* 91, 10–24.

Thornton, I. W. B. (1983). Vicariance and dispersal: confrontation or compatibility? *GeoJournal* 76, 557–564.

Tiffney, B. H. (1985). The Eocene North Atlantic land bridge: its importance in Tertiary and modern phytogeography of the northern hemisphere. *Journal of the Arnold Arboretum* 66, 243–273.

Touw, A. (1993). *Bryowijkia madagassa* sp. nov. (Musci), a second species of a genus hitherto known from continental Asia. *Bulletin du Jardin Botanique National de Belgique* 62, 453–461.

Townsend, C. H. T. (1942). *Manual of Myiology 12*. Itaquaquecetuba, Sao Paulo, Brazil.

Tozer, E. T. (1982). Marine Triassic faunas of North America: their significance for assessing plate and terrane movements. *Sonderdruck aus der Geologischen Rundschau* 71, 1077–1104.

Trueb, L., and Cannatella, D. C. (1986). Systematics, morphology, and phylogeny of genus *Pipa* (Anura: Pipidae). *Herpetologica* 42, 412–449.

Tunnicliffe, V. (1988). Biogeography and evolution of hydrothermal-vent fauna in the eastern Pacific ocean. *Proceedings of the Royal Society of London B* 233, 347–366.

Turnbull, J. W., and Griffin, A. R. (1986). The concept of provenance and its relationship to infraspecific classification in forest trees. In *Infraspecific classification of wild and cultivated plants* (ed. B. T. Styles), pp. 157–189. Clarendon Press, Oxford.

Turner, J. F. (1991). Biogeography of Australian freshwater centropagid copepods: vicariance or dispersal? *Journal of Biogeography* 18, 467–468.

Tyler, M. J. (1979). Herpetofaunal relationships of South America with Australia. *Monograph of the Museum of Natural History, University of Kansas* 7, 73–106.

Udvardy, M. D. F. (1969). *Dynamic zoogeography with special reference to land animals*. Van Nostrand Reinhold, New York.

Udvardy, M. D. F. (1975). *A classification of the biogeographical provinces of the world*. IUCN Occasional Paper No. 8. International Union for the Conservation of Nature, Morges, Switzerland.

Udvardy, M. D. F. (1981). The riddle of dispersal: dispersal theories and how they affect vicariance biogeography. In *Vicariance biogeography: a critique* (eds. G. Nelson and D. E. Rosen), pp. 6–33. Columbia University Press, New York.

Udvardy, M. D. F. (1987). The biogeographical realm Antarctica: a proposal. *Journal of the Royal Society of New Zealand* 17, 187–194.

Ulfstrand, S. (1992). Biodiversity—how to reduce its decline. Oikos 63, 3–5.

Van Balgooy, M. M. J. (1984). *Pacific plant areas 4*. Leiden, Rijksherbarium.

van Steenis, C. G. G. J. (1936). On the origins of the Malaysian mountain flora. Part 3. Analysis of floristic relationships (1st installment). *Bulletin du Jardin Botanique Buitenzorg Série 3* 14, 258–274.

Van Steenis, C. G. G. J. (1963). *Pacific plant areas 1*. Bureau of Printing, Manila.

Van Welzen, P. C., Piskant, P., and Windari, F. I. (1992). *Lepidopetalum* Blume (Sapindaceae): taxonomy, phylogeny, and historical biogeography. *Blumea* 36, 439–465.

Vane-Wright, R. I., Humphries, C. J., and Williams, P. H. (1991). What to protect? Systematics and the agony of choice. *Biological Conservation* 55, 235–254.

Vari, R. P. (1978). The terapon perches (Percoidei, Teraponidae). A cladistic analysis and taxonomic revision. *Bulletin of the American Museum of Natural History* 159.

Vari, R. P. (1989). Systematics of the neotropical characiform genus *Curimata* Bosc (Pisces, Characiformes). *Smithsonian Contributions to Zoology* 474, 1–63.

Vari, R. P. (1991). Systematics of the neotropical characiformes *Steindachnerina* Fowler (Pisces, Ostariophysi). *Smithsonian Contributions to Zoology* 507, 1–118.

Vuilleumier, F. (1978) Qu'est-ce que la biogéographie? *Compte Rendu Sommaire des Seances Societe de Biogéographie* 54, 41–46.

Vuilleumier, F., and Simberloff, D. (1980). Ecology vs history as determinants of patchy and insular distributions in high Andean birds. In *Evolutionary biology 12* (eds. M. K. Hecht, C. W. Steere, and B. Wallace), pp. 235–379. Plenum Press, New York.

Wagner, W. L., and Funk, V. A. (1995). *Hawaiian biogeography: evolution of a hotspot*. Smithsonian Institution Press, Washington, DC.

Wallace, A. R. (1855). On the law which has regulated the introduction of new species. *Annals and Magazine of Natural History* 1855, 219–232.

Wallace, A. R. 1860. On the zoological geography of the Malay Archipelago. *Proceedings of the Linnean Society. Zoology, London* 4, 173–184.

Wallace, A. R. (1876). *The geographical distribution of animals*. Macmillan and Co., London.

Wallace, A. R. (1902). *Island Life*. 3rd and revised edition. Macmillan and Co., Ltd., London.

Walter, H. (1971). *Ecology of tropical and subtropical vegetation*. Oliver and Boyd, Edinburgh.

Watson, G. F., and Littlejohn, M. J. (1985). Patterns of distribution, speciation, and vicariance biogeography of southeastern Australian amphibians. In *Biology of Australasian frogs and reptiles* (eds. G. Grigg, R. Shine, and H. Ehinann), pp. 91–97. Royal Zoological Society of New South Wales, Sydney.

Webb, P. N., and Harwood, D. M. (1993). Pliocene fossil *Nothofagus* (southern beech) from Antarctica: phytogeography, dispersal strategies, and survival in high latitude glacial-deglacial environments. NATO ASI Series A 244, 135–165.

Weimarck, H. (1934). *Monograph of the genus Cliffortia*. Kåkan Ohlsson, Lund.

Wendel, J. F., and Albert, V. A. (1992). Phylogenetics of the cotton genus (*Gossypium*): character-state weighted parsimony analysis of chloroplast–DNA restriction site data and its systematic and biogeographic implications. *Systematic Botany* 17, 115–143.

Wendel, J. F., Schnabel, A., and Seelanan, T. (1995). An unusual ribosomal DNA sequence from *Gossypium gossypioides* reveals ancient, cryptic, integenomic introgression. *Molecular Phylogenetics and Evolution* 4, 298–313.

Weston, P. H., and M. D. Crisp. (1996). Trans-Pacific biogeographic patterns in the Proteaceae. In *The origin and evolution of Pacific Island biotas, New Guinea to Polynesia: patterns and processes* (eds. A. Keast and S. E. Miller), pp. 215–232. SPB Academic Publishing, Amsterdam.

Wettstein, R. (1896). Die europaischen arten der gattung *Gentiana* aus der sektion *Endotricha* Froel. und ihr entwicklungsgeschichtlicher Zusammenanhang. *Denkschriften der Mathematisch Naturwissenschaftlichen Class der Kaiserlichen Akademie der Wissenschaften* 64, 1–74.

Wettstein, R. (1898). *Grundzüge der geographisch-morphologischen methode in der pflanzensystematik*. Gustav Fischer, Jena.

Weibes, J. T. (1982). The phylogeny of the Agaonidae (Hymenoptera, Chalcidoidea). *Netherlands Journal of Zoology* 32, 395–411.

Wickens, G. (1983). The baobab—Africa's upside-down tree. *Kew Bulletin* 37, 173–209.

Wiens, J. A. (1989). *The ecology of bird communities*, Vol 2. *Processes and variations*. Cambridge University Press, Cambridge.

Wild, H. (1952). The vegetation of Southern Rhodesian termitaria. *Rhodesia Agricultural Journal* 49, 280–92.

Wild, H. (1964). The endemic species of the Chimanimani Mountains and their significance. *Kirkia* 4, 125–157.

Wild, H. (1968). Phytogeography in south central Africa. *Kirkia* 6, 197–222.

Wiles, J. S. and Sarich, V. M. (1983). Are the Galapagos iguanas older than the Galapagos? In *Patterns of evolution in Galapagos organisms* (eds. R. I. Bowman, M. Berson, and A. E. Levington), pp. 177–186. California Academy of Science, San Francisco.

Wiley, E. O. (1976). The phylogeny and biogeography of fossil and recent gars (Actinopterygii: Lepisosteidae). *University of Kansas Museum of Natural History, Miscellaneous Publication* No. 64.

Wiley, E. O. (1977). The phylogeny and systematics of the *Fundulus nottii* species group (Teleostei: Cyprinodontidae). *Occasional Papers of the Museum of Natural History, University of Kansas* 66.

Wiley, E. O. (1981). *Phylogenetics. The theory and practice of phylogenetic systematics.* Wiley Interscience, New York.

Wiley, E. O. (1988). Vicariance biogeography. *Annual Review of Ecology and Systematics* 19, 513–542.

Williams, P. H., and Humphries, C. J. (1994). Biodiversity, taxonomic relatedness, and endemism in conservation. *Systematics Association Special Volume* 50, 269–287.

Wilson, J. B. (1991). A comparison of biogeographic models: migration, vicariance, and panbiogeography. *Global Ecology and Biogeography Letters* 1, 84–87.

Winterbottom, R. (1986). Revision and vicariance biogeography of the subfamily Congrogadinae (Pisces: Perciformes: Pseudochromidae). *Indo-Pacific Fishes* no. 9.

Wirth, W. W. (1956). Two new neotropical species of surf flies of the genus *Canace*. *Revista Brasileira de Entomologia* 5, 161–166.

Wolfe, G. W. (1985). A phylogenetic analysis of pleisiotypic hydroporine lineages with an emphasis on *Laccornis* des Gozis (Coleoptera: Dytiscidae). *Proceedings of the Academy of Natural Sciences (Philadelphia)* 337, 132–155.

Wolfe, G. W., and Roughley, R. E. (1990). A taxonomic, phylogenetic, and zoogeographic analysis of *Laccornis* Gozis (Coleoptera: Dytiscidae) with the description of Laccornini, a new tribe of Hydroporinae. *Quaestiones Entomologicae* 26, 273–354.

Wood, A. E. (1950). Porcupines, paleogeography and parallelism. *Evolution* 4, 87–98.

Wood, A. E. (1980). The earliest South American rodents: a comment. *Systematic Zoology* 29, 96.

Woods, C. A. (1982). *The history and classification of South American hystricognath rodents: reflections of the far away and long ago*. Special Publication, Pymatuning Laboratory of Ecology.

Woods, C. A. (1984). Hystricognath rodents. In *Orders and families of recent mammals of the world* (eds. S. Anderson and J. K. Jones Jr.), pp. 389–446. John Wiley and Sons, New York.

Woolley, A. R., and Garson, M. S. (1970). Petrochemical and tectonic relationship of the Malawi carbonate—alkaline province and the Lupata-Lebombo volcanics. In *African magmatism and tectonics* (eds. T. N. Clifford and I. G. Gass), pp. 237–262. Hafner Publishing Company, Darien, Connecticut.

Wormwald, N. (1984). Generating random regular graphs. *Journal of Algorithms* 5, 247–280.

Wulff, E. V. (1943). *Introduction to the historical geography of plants.* Chronica Botanica Co., Waltham, Massachusetts.

Zhao, T. Q. (1991). Fish fauna and zoogeographical division of Hexi-Alashan region, northwest China. *Acta Zoologica Sinica* 37, 153–167.

Zhengyi, W. (1983). On the significance of Pacific intercontinental discontinuity. *Annals of the Missouri Botanical Garden* 70, 577–590.

Zug, G. R. (1985). A new skink (Reptilia: Sauria: *Leiolopisma*) from Fiji. *Proceedings of the Biological Society of Washington* 98, 221–231.

Index